PLANT HUMANITIES

PLANT HUMANITIES

Edited by YOTA BATSAKI and ANATOLE TCHIKINE

DUMBARTON OAKS, TRUSTEES FOR HARVARD UNIVERSITY, WASHINGTON, D.C.

Manufactured in the United States of America
by Sheridan Books, Inc., 613 East Industrial Drive, Chelsea, Michigan 48118 | (734) 475-9145

LIBRARY OF CONGRESS CATALOGING-IN-PUBLICATION DATA

NAMES: Plant Humanities Conference (2022 : Washington, D.C.) author | Batsaki, Yota, 1974– editor | Tchikine, Anatole editor

TITLE: Plant humanities / edited by Yota Batsaki and Anatole Tchikine.

DESCRIPTION: Washington, D.C.: Dumbarton Oaks, Trustees for Harvard University, [2026] | "Volume based on papers presented at the 2022 Plant Humanities Conference held at Dumbarton Oaks, Washington, D.C., on September 15–17, 2022."—Title page verso. | Includes bibliographical references and index. | Summary: "*Plant Humanities* moves plants to the center of critical inquiry, positioning them as biocultural entities with distinct environmental and social histories that have profoundly shaped human cultures. Its thirteen chapters cover a broad geographical range, including the Americas, Europe, the Pacific, the Indian subcontinent, East Asia, and Africa, to explore the dual character of plants as placemakers and world travelers. As ecosystem builders and cultural agents constitutive of national, sacred, and domestic ecologies, plants help us trace legacies of colonialism, capitalism, racism, and the related challenges of anthropogenic climate change and biodiversity loss. Plant-focused epistemologies are also central to the volume; several essays explore the interplay between vernacular and scientific paradigms and distinct taxonomic systems to reveal instances of rupture, continuity, and resilience. The volume concludes with the discussion of the poetics of plants—their foundational role in communal imaginaries-and the ways they shape our understanding of the sacred, probing how notions of value are predicated on human commodification, elevation, or abjection of the vegetal world"—Provided by publisher.

IDENTIFIERS: LCCN 2025049308 (print) | LCCN 2025049309 (ebook) | ISBN 9780884025320 hardcover | ISBN 9780884025337 epub

SUBJECTS: LCSH: Human-plant relationships | Human ecology | Plants—Social aspects | Plants (Philosophy) | LCGFT: Conference papers and proceedings

CLASSIFICATION: LCC QK46.5.D58 P53 2026 (print) | LCC QK46.5.D58 (ebook)

LC record available at https://lccn.loc.gov/2025049308

LC ebook record available at https://lccn.loc.gov/2025049309

DESIGN AND COMPOSITION: Melissa Tandysh

MANAGING EDITOR: Sara Taylor

Volume based on papers presented at the 2022 Plant Humanities Conference held at Dumbarton Oaks, Washington, D.C., on September 15–17, 2022.

Cover, frontispiece, and section opening images reproduced from a New Zealand fern album, produced around 1883, in the collection of Dumbarton Oaks.

www.doaks.org/publications

EU GPSR Authorized Representative
LOGOS EUROPE, 9 rue Nicolas Poussin, 17000, La Rochelle, France
e-mail: contact@logoseurope.eu

CONTENTS

1

INTRODUCTION

Why Plant Humanities?

YOTA BATSAKI

Plant Humanities: Why Now?

The relationship between plants and people is deep, complex, and asymmetrical. Plants are as indispensable to humans as the air we breathe.[1] Throughout our existence as a species, we have relied on them for all our basic needs: food, shelter, clothing, and medicine. Necessary to our physical well-being, plants also weave themselves into our spiritual experience through ritual, religion, and art. We are not alone in our dependence on plants; they support almost all life on Earth through their photosynthetic alchemy and their key role in the planet's life systems. Yet because of human action, two out of five plant species are threatened with extinction.[2] Our negligence and violence toward plants are directly out of proportion with our dependence on them.

Scientists have given us all the evidence we need to understand the potentially catastrophic impact of human-driven climate change and biodiversity loss due to the burning of fossil fuels, pollution, and habitat destruction. They have calculated our diminishing window for action and offered paths to ameliorate global warming and biodiversity loss to avoid dire outcomes for our species. But the required political will and concerted action are not keeping pace with the slow violence of extinction, prompting ecological grief and even anger and despair.[3] Scientists, historians, writers, and artists are seeking new and creative ways to communicate the urgency of our predicament while reminding us of our embeddedness in the natural world. We live at a time when the pretense of a divide between nature and culture cannot be maintained. Human impact has touched the wildest places on Earth and is inscribed into the geological record for deep posterity. How can we understand and represent that impact to help effect change? Plants, long neglected in the rhetoric of conservation, offer a vital vehicle for inquiry and communication because they are integral both to our most private and local experiences and to global developments.

This is a plant moment. The resurgent interest in plants is one manifestation of what Michael Pollan has named *the botany of desire*—our longstanding manipulation of the vegetal world as an enduring feature of human cultures, shadowed by the concomitant implication that plants have shaped humans as much as we have shaped them.[4] Yet the salience of plants in contemporary scholarship is also fueled by a particularly timely current of scholarly attention to more-than-human beings as historical agents. Plants are prime candidates for this kind of inquiry due to their outsize imprint on several core features of modernity. Over the past half-century, historians have documented the transformations brought about by early modern plant transfers, colonial competition, and the rise of the plantation complex as a foundational practice of capitalism and empire.[5] These extractive practices centered on plants—and their cultivation and commodification at scale—shaped modernity alongside the industrialization powered by fossil fuels. Their legacies include accelerating species loss, a changing climate, and the spread of invasive species that render plants crucial indices of change, collapse, and resilience. Taken together, they go a long way toward illuminating our current social and environmental predicament.

Beginning in the early modern period, the influx of global specimens into European collections, the networks of botanical gardens as laboratories for experimentation, the vast archives they engendered, and the plantation system that grew out of these developments triggered vast voluntary and forced migrations. Sessile, plants travel constantly through their adaptations for nourishment and reproduction. From the sixteenth century onward, this movement occurred in tandem with the massive displacement of people. European interest in the world's floras exploded as imperial competition accelerated. The increased traffic of specimens through imperial voyages of exploration and trade brought the challenge of identification and classification and breakthroughs in taxonomy.[6] As the movement and spread of plants grew larger and faster around the globe, their historical impact was also magnified by desire and greed. The historiography of these migrations has grown over the past fifty years beyond the Columbian Exchange to acknowledge the botanical contributions of enslaved people of the African diaspora and the erased or appropriated knowledge of Indigenous communities.[7] This growing recognition of the mobilization of plants as a motor of historical transformation partly accounts for their current prominence and animates several chapters in this volume.

Botany in the eighteenth and nineteenth centuries was big business.[8] The new practices and technologies that underpinned the scientific study of plants (herbarium specimens, botanical illustrations and publications, botanical gardens) extracted plants from their local ecosystems and, therefore, from their relationships to other species and the knowledge traditions and landscape management practices of local human communities.[9] Plants often thrive away from their native habitats due to the absence of pests and disease. The transplantation of economically significant plants yielded tremendous profits. Coffee grows better in South America than in Africa; rubber prospered in tropical Asia rather than in the Amazon. But the decontextualization of plants that enabled their scaled-up cultivation in plantations brought a terrible human toll: the transatlantic slave trade and the degradation of ecosystems in the Caribbean and the Americas through monocultures of sugarcane, cotton, and tobacco. Biopiracy also fueled colonial expansion: the theft of cinchona seeds from Peru secured the antimalarial quinine that supported British imperialism in West Africa and India.[10] The dislocations that followed transformed ecosystems, destroyed habitats, and precipitated the rapid human alterations to the planet's life systems that we experience as the Great Acceleration. Scholars have given another name to the Anthropocene to drive home the link between colonialism, the violence of the plantation system, the environmental toll of monoculture, and their legacies of environmental and social injustice: the Plantationocene.[11] Several chapters in this volume analyze global plant commodities, such as tea or oilseeds, situating them at the center of ecological and social transformations whose effects linger today.

FIGURE 1.1
Barthélémy Toguo, *Homo Planta I*, 2018. Ink on canvas, 100 × 100 cm. © Barthélémy Toguo / courtesy of Galerie Lelong & Co and Bandjoun Station.

In the humanities, plant thinkers over the past few decades have also worked to reclaim plants from theoretical neglect. Earlier interventions by scholars like Val Plumwood and Carolyn Merchant paved the way for a critique of the objectification of plants as passive resources by showing how narratives of scientific progress and economic "improvement" were premised on notions of nature as a feminized, passive entity.[12] The relegation of plants to the margins of critical thought was intensified by Western scientific emphasis on taxonomic control of the natural world, abetted by philosophical assumptions about their lack of sentience and agency. Within the burgeoning field of critical plant studies, scholars like Matthew Hall, Michael Marder, and Jeffrey Nealon point to traditions within Western metaphysics that have relegated plants to the bottom of the scale of being due to their lack of a face, their seeming immobility, or their apparent captivity by their environment.[13]

Another perspective has come from culturally and geographically specific histories of plants that serve as windows to global circuits of power and empire. By focusing on nutmeg, for instance, Amitav Ghosh tied together early modern practices of settler colonialism—the subjugation or extermination of Indigenous populations, coupled with the transformation of local landscapes and ecologies—to a vision of the "world-as-resource" that serves a limitless extractive logic.[14] The plantation complex, as the epicenter of forced plant-human mobilities, combines the objectification of plants and the abjection of humans. The porous boundary between human and more-than-human worlds, shaped by violence and sustained by resilience and hope, finds its embodiment in the fused being of Barthélémy Toguo's *Homo Planta I*, an image that transmutes the suffering of humans and nature into an intimation of new energies about to sprout (Figure 1.1). Inherent in this dismantling of the nature-culture opposition is the recognition of inescapable human entanglements with other organisms and the urge to reimagine those relationships, whether in the precarious assemblages of the patchy Anthropocene proposed by Donna Haraway and Anna Tsing, or in the restoration of reciprocity with more-than-human beings advocated by Robin Wall Kimmerer.[15] And so, the current historical trend of centering the plant as a historical actor is also a planetary ethical imperative: "the task of imaginatively restoring agency and voice to nonhumans . . . a task that is at once aesthetic and political."[16]

Within the current efflorescence of plant-focused modes of scholarly inquiry, why the need for a "plant humanities"? While the term is recent, coined for the Plant Humanities Initiative at Dumbarton Oaks that was launched in 2018, the space it maps out is not entirely new.[17] The study of how plants have shaped human cultures traverses ground that goes back millennia, from Indigenous practices and knowledges to the social transformations brought about by plant domestication, to colonial botany and the rise of racialized plantation systems and industrialized agriculture. Ethnobotany and economic botany were, perhaps, the first to map out this space, joined more recently by the environmental humanities, ecofeminism and ecocriticism, and critical plant studies. The plant humanities are in dialogue with these methodologies and yet bring distinct characteristics of their own. First, the environmental humanities are vast in scope. The plant humanities share an interest in urgent questions of biodiversity loss, climate change, and their unequal impact on human and more-than-human beings. But they center their inquiry on the vegetal world as a distinct sphere of biocultural inquiry with far-reaching historical, political, and ethical implications. Moreover, the methodology of the plant humanities often eschews the abstraction of vegetal life into a theoretical concept, concentrating instead on a particular species or plant family to tease out their distinctive evolutionary and cultural histories. Several chapters in this volume highlight specific species and their entanglement with local ecologies and communities.

Another characteristic of the plant humanities is a focus on plant collections as fertile ground for interdisciplinary inquiry. The historical processes that transformed the planet's human and more-than-human communities have left their record in the massive plant archives that reside in the world's natural history collections. These biocultural repositories provide incomparable material for examining how knowledge about plants was acquired, organized, circulated, erased, and exploited over time.[18] Several chapters in this volume hone in on the herbarium, the practice of preserving collections of dried plants on paper, whose roots in early modern natural history are often obscured by their later colonial entwinements. Herbaria that were first conceived as resources for the early modern teaching of medicine, as tools of scientific taxonomy, and later as engines of economic botany are now being used to reconstitute past human-plant communities or to model the effects of climate change.[19] While their colonial hauntings are receiving deserved scrutiny, the recent reimagining of herbaria as biocultural collections opens them up to new uses by scientists, humanists, and artists alike. They are versatile technologies or "exaptations" that afford new meanings, connections, and opportunities for ethical engagement.[20]

Such finely honed encounters with plants in place, on the move, or in the archive—attuned to the particularities of time, place, and organism—are urgently relevant to broader humanities questions. Through them, the plant humanities partake in the larger historical and philosophical effort to rethink our connection to more-than-human beings and reclaim them as historical agents. If developments over the past few centuries have conspired to instrumentalize plants, then rekindling our awareness of the different kinds of value, beyond the economic, that have been attached to them across human societies may help restore intellectual depth and ethical significance to plant-human relationships.

Plants and Human Affairs

Plants have played a central role in human evolution and culture from our very beginning. It has been suggested that the spread of flowering plants provided "a springboard for primate evolution."[21] The utility of plants was not limited to food but encompassed the whole spectrum of human needs. They provided weapons, dyes, and cordage. They were fashioned as garlands for the dead; distilled into embalming liquids and perfumes for ritual; and supplied medicine, poisons, and hallucinogens. Calendar plants functioned as seasonal markers that shaped our ancestors' sense of time and the rhythms of labor. Prehistoric and traditional communities had intimate knowledge of plants that was woven into every gesture and area of human activity. Although we decry our "plant blindness,"[22] we know that many Indigenous classification systems record hundreds of species and their properties, so obliviousness to plants is not necessarily a cognitive but an epistemological predicament.[23] "One thing of which we can be sure is that more of this knowledge is lost than is now known."[24] This recognition is nowadays expressed in renewed respect for Indigenous knowledge and increased sensitivity to the ethics of ethnobotanical research.

From our early education, most of us learn of the pivotal development of agriculture ten to twelve thousand years ago. Emerging independently in different places (i.e., rice in Asia, wheat and barley in the Middle East, rice and pearl millet in West Africa, maize in Mesoamerica), plant domestication led to sedentary communities, social stratification, population growth, and environmental degradation—but also to surpluses and leisure for scientific and artistic pursuits. We are taught that domestication brought plants under our control. But we might as well conceive of plants enlisting humans alongside their manifold strategies for survival, travel, and reproduction, inducing us to tend to and propagate them and shaping our societies and lives in the process.[25]

We rely on plants to not only nourish but also heal our bodies. *Materia medica*, or texts detailing the medicinal uses of (mostly) plants, is one of the oldest recorded traditions of natural history. The first-century CE Greek scholar Dioscorides was a key source for the transmission of East Mediterranean knowledge about the medicinal uses of plants to Europe well into the sixteenth century. Older traditions of *materia medica* survive in archaeological findings from Mesopotamia and records from Asia (including China, Korea, and India), while the traditional pharmacopoeia of Mesoamerica, Africa, and the Balkans are of keen interest to today's ethnobotanists.[26] In these traditions, the culinary, medicinal, and spiritual uses of plants were often intermingled, belying the separation that followed the rise of Western science. Much traditional knowledge about the medicinal virtues of plants has been lost or ignored, and we remain unaware of the potential medicinal value of more than 90 percent of the plants known to us, let alone species not yet discovered. Ethnobotanists today often turn to these older traditions in search of plant compounds that will inform pharmaceutical breakthroughs in the future. Because of past biopiracy, such research raises ethical as well as legal and financial questions. Meanwhile, the World Health Organization (WHO) estimates that 80 percent of the world's population uses herbal medicine and millions in the Global South still depend on plants for their primary health care.[27] Deforestation and loss of biodiversity pose an immediate threat to the well-being of these communities.

While plants have been our chief resource for healing, they can also cause suffering: "the four most popular addictive substances—sugar, caffeine, tobacco, and alcohol—are all derived from plants."[28] The story of Western imperialism in China can be spun around two addictive plant substances. The Opium Wars were born out of Britain's desire to redress the trade deficit caused by the consumption of Chinese tea, among other commodities, by exporting Indian opium to China. Although human violence and greed lie at the root of these narratives, they are also reminders of the power of plants over our bodies and minds. As a source of poisons as well as remedies, the photosynthetic world frustrates our attempts to contain its effects. While we may revere plants as models of benevolent coexistence and reciprocity, their behavior is often competitive and deleterious to other organisms, including plants and humans. Taking the fullest measure of the vegetal world's diversity and of our differing modes of historical and sensory interaction requires us to negotiate, as several chapters in this volume do, the limits of human language, metaphors, and access to alternative modes of being.

Necessary to all our bodily needs—breath, nourishment, healing, and pleasure—plants are perhaps the most resonant vehicle for reflecting on our embodied experience of the world. Symbolically, they are markers of important transitions. The fruit of the tree of the knowledge of good and evil in the Christian tradition mediates Adam and Eve's passage into a moral, bodily experience of labor and need. The Buddha found Enlightenment under a bodhi tree. Indigenous communities in South Africa and North America alike burn herbs—impepho and sage—to purify space, bridge different temporalities, and establish communication with the ancestors. The Bronze Age peoples of the Eastern Mediterranean engaged in long-distance trade of complex aromatic substances extracted from flowers, "oils of gladness" that symbolically distilled the whole of nature and were used in celebrations of spring rebirth and rituals of transition: from birth, to marriage, to death.[29] This symbolic significance of flowers survives in folklore, the material culture of the past, literature, and art. And, as the two concluding chapters of this volume remind us, the intimate link between plants and the sacred offers a neglected but potent reminder of their value, and perhaps even ground for adjudicating their legal and ethical standing.

Conversely, how can our investigations of plant histories enrich our ethical response to human communities? Maize is a perfect example.[30] We owe the domestication of maize to the ingenuity and labor of the Indigenous people of the Americas, many of whom called themselves "maize people." Linguistic and archaeological evidence reveals a history of migration. Initially domesticated in Mesoamerica, maize migrated four thousand years ago to South America, where it was diversified and enhanced as a domesticated species in Amazonia before being reintroduced to Mesoamerica. The Indigenous societies of the Americas developed the complex processing techniques that enhanced the nutritional value of maize and contributed to the flourishing of local civilizations before the plant's migration to the rest of the world. Maize, as we know it today, came about through a symbiotic relationship with humans.

Today, this plant-human relationship is severed by large corporations that genetically manipulate maize so that its seed cannot be resown. Historical inquiry into how maize has coevolved with humans over the past six thousand years shows that our scientific and industrial systems are indebted to Indigenous knowledge. Yet globalization and its practices of genetic copyright and mass production adversely impact the very communities where the knowledge and domestication of the plant were developed in the first place. Maize was not only a source of nutrition for the communities that made it available for human consumption. It was central to Maya cosmology; their creation story in the *Popol Vuh* recounts how humans were fashioned by the deities out of maize. This symbolic and ancestral attachment to the plant galvanizes popular protests, such as the 2014 demonstrations in Guatemala against the so-called Monsanto Law, which sought to prohibit the replanting or transportation of privatized seeds, effectually ending further experimentation and severing the symbiotic bond between the plant and the community.[31] Plants are intertwined with human communities through food traditions, land practices, rituals, and labor. The example of maize suggests that current debates about food security or environmental justice cannot dispense with historical awareness of plant-human relationships.

Acknowledging plants as a shaping force of human cultures requires interdisciplinary conversations that can be rooted in the insights of science, paying heed to the specificity of organisms, their properties, environments, and evolutionary trajectories, and that, at the same time, draw on humanistic modes of inquiry to narrate and interpret our present and imagine the future of multispecies relationships of which we are necessarily a part. Ambitious undertakings such as the Earth BioGenome Project may benefit from collaboration with humanists to communicate their aims and wrestle with the ethical questions raised by the effort to sequence the genomes of all living species.[32] The project's aims of "[i]ncreasing our understanding of Earth's biodiversity and responsibly stewarding its resources are among the most crucial scientific and social challenges of the new millennium." The social challenges include reaching an agreement on the best way to safeguard and share information about biodiversity that may inhere in Indigenous communities and localities that have historically experienced biopiracy or unequal sharing of their resources. An open-access framework may be the most conducive to scientific investigation and sharing, yet it may not benefit the Indigenous peoples and local communities that have developed and transmitted knowledge about local biodiversity for centuries. At the same time, a humanistic reading of the nation-state as an "imagined community" predicated on constructed narratives of autochthony and purity may function as a corrective to a narrow proprietary approach to resources that may benefit humanity as a whole and safeguard a multiplicity of species. Questions of who stands to benefit from biodiversity today cannot be answered without analyzing how knowledge was circulated and appropriated in the past, nor can they be resolved without ethical inquiry.

Scientific accounts of plant evolution, when combined with humanistic and aesthetic approaches, offer rich scope for ecological and ethical awakening. Consider how evolution, aesthetics, and ethics converge in the case of a tropical beauty like heliconia, which originates mostly in Central and South America but has spread as an ornamental plant around the globe. This work by botanical artist Margaret Mee, in the Rare Book Collection at Dumbarton Oaks, celebrates the beauty of the plant through an accuracy of depiction that appeals to botanists and a wondrous use of color and composition that enthralls art connoisseurs (Figure 1.2). It exemplifies a key convention of Western botanical art and scientific illustration, rendering the singular specimen removed from its natural context and portrayed against a neutral background. The emphasis is on those attributes of the plant, such as leaves, flowers, and reproductive organs, that aid its identification and classification. Similarly, the herbarium sheet removes the plant from its ecosystem, draining it of color, scent, and texture, then stabilizing and rendering it suitable for transportation, archiving, and future reference in a natural history collection. We *know* that plants do not exist in isolation, that they inhabit diverse habitats populated by other plants, animals, and microorganisms. Yet the visual technologies that mediate our engagement with the natural world often paradoxically induce a certain kind of blindness. Developed to transmit knowledge and pleasure, they also entail information loss.

Two etchings by Bryan Poole, based on his own experience in the field with botanist W. John Kress, re-embed the plant in mutualistic relationships

FIGURE 1.2
Margaret Mee, *Heliconia, Proc: Amazonas, Nr. Manaus, November 1964, Margaret Mee*, 1964. Gouache, 66 × 48 cm. Rare Book Collection, Dumbarton Oaks, Washington, D.C.

FIGURE 1.3
Bryan Poole, *Heliconia bihai*, 2008. Colored etching, 75 × 55 cm. National Museum of Natural History, Smithsonian Institution, Washington, D.C.

FIGURE 1.4
Bryan Poole, *Heliconia caribaea*, 2008. Colored etching, 75 × 55 cm. National Museum of Natural History, Smithsonian Institution, Washington, D.C.

and document a striking example of coevolution between plant and pollinator (Figures 1.3 and 1.4). The male and female of a single species of hummingbird (*Eulampis jugularis*) on the Caribbean island of Dominica have evolved to visit different species of heliconia. The males, characterized by short, straight bills, only visit the compact flowers of *Heliconia caribaea*; by contrast, the females, equipped with longer curved bills, prefer the elongated flowers of *Heliconia bihai*. Such elaborate mutualistic relationships evolved in deep time within ecosystems that are increasingly under threat. The characteristics that made the plant attractive to gardeners and artists—its vibrant color and sculptural form—are also markers of its coevolution with the hummingbirds that pollinate it. The bright reds and yellows attract the hummingbirds, while the various shapes of the bracts have evolved to discriminate among and guide the preferred pollinators. Understanding how form serves reproductive function situates the plant in a web of relationships in which humans, as destroyers or stewards of the ecosystems where these relationships evolved, play a key role.

Going beyond the heliconia's appearance to grasp the complex biological mechanisms by which a species may recognize and discriminate in favor of a preferred hummingbird species unsettles our notions of cognition and decision-making without a need to apply anthropomorphic notions of agency to plants.[33] Rather, we may walk away from such mutualistic networks with an expanded notion of what cognition, movement, or action can mean across radically different modes of being.

Contrary to the poet John Keats's complaint that science unweaves the rainbow, science can ignite our wonder and appreciation; when combined with engaging storytelling, it sharpens our empathy for the beauty and fragility of these unique evolutionary relationships of which we partake.[34] This work of synthesis across disciplinary boundaries animates the plant humanities as they were conceived and are being cultivated at Dumbarton Oaks and at other centers of scientific and humanistic inquiry into plants, often converging around significant plant collections.

Under the umbrella of the plant humanities, scientists, humanists, and artists can join forces to interrogate the modern economistic view of plants. Even well-intentioned conservation attempts based on "ecosystem services" apply a narrow cost-benefit analysis to the natural world. Plants are valued for their quantifiable contributions to the well-being of other species or the productivity of landscapes. And yet, "we have tried using the utilitarian language of economics in order to protect nature for more than a hundred years." As Lyanda Lynn Haupt points out, "It has not worked."[35] The humanities, by contrast, privilege experience and draw on qualitative measures for assessing value and meaning. They explore how relationships of reciprocity and care enrich our existence and incite a duty of stewardship that may well exceed our lifetime or historical moment. To care for climate change or biodiversity loss, we need to feel moved by interests that are not narrowly our own and a future that we may not experience—in other words, to adopt a time frame very different from current political and economic cycles and the calculations they engender.

Perhaps no category imposed on the vegetal world yields more insight into the contingency of values attached to plants than that of "invasives." Kudzu (*Pueraria lobata*) is reviled as the "vine that ate the American South," but, in its native China, Japan, and Korea, it has been prized for millennia as a medicinal and useful plant and an antidote to rural famines.[36] It crossed the ocean to America as an ornamental for Southern porches, before being tried out as fodder and later massively deployed by Franklin Delano Roosevelt's New Deal to combat soil erosion in the former Cotton Belt. While the military metaphors of invasion biology cast invasives as harmful aliens and colonizers, a humanistic perspective might find in kudzu a guide to the environmental history of settler colonialism and plantations in the South. The reasons for its proliferation highlight our always partial and situated scientific knowledge of plants—in the 1940s, warnings about the plant's potential invasiveness were dismissed, while today similar warnings may be exaggerated due to the vine's hypervisibility on the embankments and roadsides where it was planted to retain the soil. At the same time, eradication strategies seem to be motivated not only by kudzu's negative impact on biodiversity (which, in the case of this, as many other invasives, is not fully understood) but also by the vine's negative impact on timber plantations, a new form of monoculture in the ravaged landscapes of the South.

Contemporary artists are alert to the generative archive of plant histories and its potential for resurrecting the past, grasping the present, and imagining alternative futures. Kudzu was the protagonist of a massive installation in a central location of the 2022 Venice Biennale, populating an interior, invasive garden also inhabited by sugarcane and various critters. The installation by Precious Okoyomon was entitled *To See the Earth Before the End of the World* (Figure 1.5). Spread among the kudzu were brown wool sculptures encircled by the growing vines. Okoyomon's garden, with its telling inclusion of another plantation plant, sugarcane, alludes to kudzu's ubiquity in the South as the uncanny successor of cotton. The sculptures in the kudzu suggest the toiling bodies on the plantation, but their gestures are not those of toil; they gracefully gesture forward or sink into the vines in a poignant and subtle memorialization. The artist's disruptive gesture of bringing an invasive living organism into the gallery and allowing it, through its rampant growth, to co-produce the installation exemplifies those powerful currents in contemporary aesthetic and ethical practice that seek to restore animation and agency to more-than-human entities. The beings brought together in *To See the Earth Before the End of the World*—kudzu, sugarcane, black swallowtail butterflies, human visitors, and sundry soil inhabitants—are not elements of

FIGURE 1.5
Precious Okoyomon, installation view of *To See the Earth Before the End of the World*, 2022, at the 59th International Art Exhibition, Venice Biennale, "The Milk of Dreams," 2022. Photograph by Clelia Cadamuro / courtesy of Precious Okoyomon.

a native ecosystem. They are part and parcel of the migrations and dislocations that have shaped the Anthropocene, coming together in an aesthetic assemblage that suggests new "ruderal ecologies" and forms of coexistence and belonging.[37]

A similar critical assessment applies to attempts at the restoration of past ecosystems; scientists of evolution accustomed to thinking in deep time may point out the arbitrariness of choosing a particular epoch, given that floras have changed and moved with periodic glaciations and the movement of the continents. Moreover, while it might seem self-evident and benign to campaign for a return to ecosystems putatively untouched by humans, historians and theorists have situated the concepts of "native" plants within troubling ideologies of racial and cultural purity.[38] The militaristic rhetoric of invasion biology is akin, in this reading, to the xenophobic rejection of human migrants as alien intruders. While such notions of purity bolster the "imagined community" of the nation as static, rooted, and homogeneous, our histories and present are shaped by the irresistible flux and constant migration of organisms—plants, pathogens, and humans.[39]

Chapter Summaries

The volume is divided into five interrelated sections. The first two—Mobilities and Emplacements—explore some of the ways plants have shaped human histories through their dual capacities for migrating and placemaking. The terraforming system of the plantation as a nexus of plant and human mobilities ties together several chapters. The third—Taxonomies—explores human attempts to impose order on the world of plants and our ways of knowing them through our bodies and our epistemologies, with an emphasis on the herbarium as the paradigmatic plant archive. Fourth is a section on Poetics that extends these ways of knowing to the play of language and signification, probing how the silent opacity of plants serves both to map and undo human and more-than-human ontologies. Lastly, a section on the Sacred reminds us of the past and ongoing spiritual dependence of humans on plants, in order to restore to them a value that goes beyond narrow profit and use.

The first section, Mobilities, works against the grain of our common perception of plants as rooted to a spot. Far from being prisoners of their environment, plants travel the world through their adaptations; in the modern period, their travels have set into motion massive voluntary and involuntary human displacement. While these stories of plant-human mobility inscribe colonial violence and the sinister transformations of the Plantationocene, they also speak of resilience, refuge, and new forms of flourishing. Judith A. Carney and Richard N. Rosomoff have established the botanical agency of enslaved peoples from Africa within the narrative of the Columbian Exchange first advanced by Alfred W. Crosby. Their chapter on "African Foods in Diaspora: Sustenance, Worldmaking, and Neotropical Legacies" shows how the adoption of African food plants in the Americas is deeply intertwined with the history of New World slavery. Forcibly transported in vast numbers during the 350-year Atlantic slave trade, African people brought seeds, farming skills, and culinary traditions to plantation societies in the Caribbean, Brazil, and the United States. The story of how important crops such as African rice (*Oryza glaberrima*) became established in the Americas exemplifies the botanical knowledge, agency, and heritage of the enslaved and their descendants. Their self-provisioning strategies—from slave ship stores of resilient seeds to small-scale gardens in the margins of plantations to the "escape agriculture" of Maroon communities in places like Suriname, Bahia, and Maranhão—were not only key liberatory practices but also sources of rich and enduring repositories of agroecological knowledge: "biocultural refugia" that remain vital today as a counterpart to monocultures.

In "Palm Readings: Plantations, Colonialism, and Racial Capitalism in Indonesia and Mexico," Perdana P. Roswaldy and Jayson M. Porter offer another perspective on the Plantationocene through the trajectories of two important oilseeds—coconut and palm oil—traveling through global commodity chains shaped by imperialism and racial capitalism. The coconut palm (*Cocos nucifera*), native to both Southeast Asia and the Americas, became a linchpin of colonial economies, circulating via ancient human migrations, colonial trade, and even slave ships. In Indonesia, coconut production led to the supplanting of Indigenous agrobiodiverse plots by racialized land and labor regimes. In Mexico, coconuts arrived through a blend of precolonial dispersal, Spanish colonial trade, and later state and foreign investment. There, too, they replaced biodiversity-rich landscapes, while plantations enforced new orders of exploitation, labor segmentation, and land dispossession. Roswaldy and Porter demonstrate how the mobility of the coconut—and later, oil palm (*Elaeis guineensis*)—tied together global networks of capital, labor, and knowledge. The plantation emerges from these two case studies as an adaptable and enduring institution whose aftermath continues to define the contours of global capitalism, resource extraction, and local resistance.

Beyond the historical circuits of empire and its aftermath, the plant world provides some of our most potent metaphors for the violence and promise of transplantation. In "Migrations: Tales of Uprooting, Transplanting, and Grafting," Giovanni Aloi weaves together several intimate encounters with plants—living, dried, and remembered—to highlight how human and more-than-human travels are enmeshed with memory, family, and emotion. Whether in the heavenly scent of a lemon grove in flower in a suburban garage gripped by the winter cold; the transmission of a zimmerlinde (*Sparrmannia africana*) through generations of the displaced Freud family; or the receipt of a box of succulents gifted through the author's maternal lineage, plants serve as living repositories of personal and collective memory across geographies and generations. The act of propagating, preserving, and transmitting plants—as cuttings, relics, or grafts—mirrors the uprooting and transplantation experienced by human migrants, marking the shifting boundaries of identity and the vagaries of taking root in new soil. Artistic interventions, from Otobong Nkanga's *Diaspore*

(2014) to Fatma Bucak's grafted *Damascus Rose* (2016), turn to plants to transmute the pain and hope of the foreigner, the exile, and the refugee. By acknowledging our kinship with plants as fellow travelers in time and space, Aloi argues, we might just recover our precarious foothold on belonging and make the graft take.

Any account of the dynamic and multivalent mobility of plants must also pay attention to their unique placemaking ability, their radical conjuring of the emplacements of human and more-than-human beings. Plants are the chief ecosystem builders, and their presence sustains and often determines myriad other interspecies relationships and interactions. At the same time, any attempt to redress the habitual neglect of plants by placing them at the center of scholarly inquiry runs the risk of obscuring the reality of their interdependence with other plants and animals and decontextualizing them anew. Heeding this fundamental interdependence and its conceptual and ethical implications, Romita Ray asks: "As the emerging discipline of the Plant Humanities engages with how plants both shape and are a part of human life worlds, can it make room for the animal interlocutors with whom the material and affective dynamics of plant and human lives remain enmeshed?" In "'Joyous Howling': Hoolock Gibbons, Tea, and Plantation Landscapes," Ray explores how the transformation of India's jungles into tea plantations radically reshaped place and multispecies relationships by tracing the interwoven trajectories of the hoolock gibbon—a loud but mostly invisible presence in the dense forest—and the British discovery of Assam *jungli* tea (*Camellia sinensis*). As the British imperial project of tea cultivation at scale invaded local ecosystems, both plants and animals were extracted and reconstituted as scientific specimens, instruments of economic "improvement," and domesticated companions, their dense habitats replaced by plantation landscapes engineered for profit and order. The precarious survival of creatures like the hoolock gibbon in the midst of these radical dislocations reminds us that the shifting border with the wild and untamed—the jungle as, in Ray's phrase, "the very antithesis of the modern"—actively shaped plant capitalism, drawing the dynamics of tea (a key plant commodity) into a wider, ever-shifting web of interspecies relations that both propelled and unsettled the expanding plantation landscape.

While the tea plant is a staple of economic botany and a major historical player in colonial and plantation histories, Rosetta S. Elkin's chapter turns our attention to plants that "persist along the margins of human life, making history scrappy, dirty, or noisy, because they have no immediate worth or use." By focusing on a plant that thrives in marginal landscapes, "Blowout Beardtongue: The Glitch in Landscape Conservation" offers a critique of misguided attempts to redress the harm of deforestation, industrial agriculture, and the plantation complex. The case of blowout beardtongue (*Penstemon haydenii*) reveals the limits of such human-driven interventions. This rare, federally protected plant flourishes only in the unpredictable, wind-scoured depressions called blowouts, disturbed habitats that resist stabilization in the Nebraska Sand Hills. While a focus on productivity construes instability as a problem, blowout beardtongue's strategy of latency—sustaining itself through abundant yet hidden seed banks—embraces unpredictability and change. The plant germinates and spreads inscrutably on its own terms, thwarting attempts to domesticate its life cycle and reminding us that "scrappiness" and mutability may be sources of resilience. By centering the unruly agency of plants, the chapter demonstrates how well-meaning restoration projects often misunderstand the complexity and autonomy of natural systems, suggesting that the preservation of entire habitats, rather than individual species, should be the focus of attention and care.

Taxonomies are systems of order that also double as hierarchies of value and as vehicles for establishing or severing relationships. While we usually look for them in scientific textbooks and natural history museums, Anatole Tchikine urges us to take a more expansive and adventurous view of the plant archive. His chapter focuses on food taxonomies gleaned from a work of twentieth-century juvenile literature, *The Adventures of Cipollino*. Throughout the evolution of Western botanical science, traditional and vernacular forms of plant knowledge, Tchikine argues, have not disappeared but were instead overlaid by new systems of classification, from medieval cosmologies and household hierarchies to the influential structures of Linnaean taxonomy and modern phylogenetics. Cipollino's world of personified fruits and vegetables is infused with the echoes of medieval and folk hierarchies, a reminder of how everyday culinary practices encode a social order of plants that is intelligible and meaningful outside scientific norms. By contrast, the work of the sixteenth-century botanist Andrea Cesalpino is invoked to show how the rise of scientific systems often entailed the exclusion of culinary, medicinal, and contextual plant knowledge—categories that nevertheless persisted in household traditions, recipes, and local markets. Thus, rather than erasure, the history of plant classification reveals an ongoing layering, with food rituals acting as a living record of overlapping taxonomies. Tchikine calls on the plant humanities to engage with long-standing dietary and herbalist traditions—from annotated herbals to recipe books and agricultural treatises—that "often bear an imprint of non-elite practices" and can, therefore, enrich our appreciation of Indigenous and vernacular relationships with plants.

If it is not living food cultures, but rather desiccated specimens catalogued in scientific cabinets, that usually serve as the privileged site of plant taxonomy, then Ashanti Shih encourages us to reclaim the herbarium as "a site for inquiry in the plant humanities," to uncover its erased and forgotten histories. Her foray into the "Herbarium as Living Archive" began with a type specimen of ʻeke silversword (*Argyroxiphium caliginis*)—an extremely rare plant existing in a single volcanic crater on Maui—and the erasure of a name from its herbarium sheet. Set in early twentieth-century Hawaiʻi, the chapter traces the collaboration between Henry Wiebke, a Kanaka Maoli (Native Hawaiian) botanist, and Otto Degener, a German American botanist, who together collected thousands of Hawaiian plant specimens, including the ʻeke silversword, only for Wiebke's name to be systematically struck out from the record. This pattern of erasure was not limited to Wiebke but also included Asian collaborators whose lives were entwined with Hawaiʻi's plantation economy. Degener's acclaim demonstrates how systems of power in science have historically privileged white settler-colonial actors and silenced local and Indigenous voices. At the same time, Shih proposes that herbaria are replete with counter-histories of collaboration, resistance, and Indigenous relationality to land. By urging humanists to engage critically with these "living archives" and their "colonial hauntings," Shih argues for a decolonial reading of scientific collections that can confront past injustices and open possibilities for future recognition and repair.

In a similar vein, Yota Batsaki's chapter, "Disruptive Taxonomies: In the Herbarium with Temitayo Ogunbiyi," built around the contemporary Nigeria-based artist Temitayo Ogunbiyi's series of drawings on herbarium paper, explores the generative potential of the herbarium through aesthetic engagement with this foundational practice of early modern natural history. In interrogating the herbarium as a record of loss, the chapter engages critically with Michel Foucault's account of dried plant specimens as paradigmatic of the "Age of Classifying," wherein beings were "stripped naked" of context, stories, and local relationships in the service of Western classification systems. Foucault's influential account risks reinforcing a static, lifeless view of archives and natural history collections that exist, instead, on a continuum of liveliness (from botanical gardens to seed banks to herbaria) and are not fully determined by the historical conditions of their production. As a counterpoint, the chapter turns to Ogunbiyi, whose residency in the United States National Herbarium inspired her to work on herbarium paper. The artist creates botanical drawings and paintings of plants—such as okra, pineapple, or cocoyam—that resonate with social meanings and are accompanied by handwritten annotations that re-embed them in culture and the imagination. Viewed as a case of what Jacques Derrida calls "archive fever," Ogunbiyi's practice resists the herbarium's originary violence and dislocation: she restores the plant's aliveness and indeterminacy and reinterprets the herbarium sheet as a site of ongoing contestation, creativity, and promise.[40]

Lucas Mertehikian's "The Quiet Science? Jorge Luis Borges's Anti-Botany" extends the interrogations of plant taxonomies in the previous chapters to probe how the perceived silence and immobility of the vegetal world test the very contours of the human. The chapter forms a bridge between the previous section on Taxonomies and this one on Poetics, pursuing questions of ontology and indeterminacy brought out by Ogunbiyi's visual practice but reconfiguring them through the understated presence of plants in the works of Argentine writer Jorge Luis Borges. Borges's description of botany as the "quiet" science aligns with Western philosophical traditions that cast plants as passive entities relegated to the periphery of thought. Through an attentive reading of the short story "Funes, His Memory" in the broader context of Borges's writing, Mertehikian draws our attention to the similarly quiet but radically disruptive potentiality of plants in the literary imagination. The main character, Funes, is immobilized after an accident but acquires extraordinary powers of perception and memory. The narrative dwells on Funes's subsequent and intense contemplation of plants—such as the fig tree, artemisia, and, most notably, the passionflower (*Passiflora*)—in all their vibrant and seemingly inexhaustible variety. Neither scientific nomenclature nor common names can contain the experience of Funes's plant-human sensorium. In challenging the perceived marginalization of plants in Borges's work, Mertehikian argues more broadly for a poetics that, in its attunement to the "otherness" of plants, affords the possibility of enlarging our sense of human subjectivity and difference, groping toward less bounded notions of being.

But is it possible to access this "otherness" in ways that might inform our own intellectual and ethical responses to plants? What forms of communication would such an endeavor at cross-species connection entail? Sumana Roy's poetic meditation, "'Phooler Bhasha': Rabindranath Tagore's Flower Linguistics" explores the idea of a "plant script"—the possibility that the secret being of plants can be transcribed and decoded in the manner of human language. Roy references the work of Jagadish Chandra Bose, who sought to record plants' responses to stimuli through scientific instruments, before moving to Rabindranath Tagore's fascination with the "language of flowers." Through poetry and song, Tagore strove to capture the powerful emotional and sensory hold of the flower's fragrance, its exuberant blooming, and poignant ephemerality. For Tagore, the language of plants is suffused with longing and beauty yet constantly elides attempts at translation into human words: the exuberant flower instantiates a conversation with the world, and its fall provokes an ineffable yearning and loss. Roy reflects on her own attempts to create a script of leaf impressions—a form of nature printing—only to decide that this method only increases her distance from the plant's unique expressiveness. Ultimately, it is "traveling in the opposite direction," registering the significance of the flower not as ornament but as a living necessity, that reanimates poetic rhetoric and yields glimpses into a "shared grammar of existence."

The question of poetic relationality to more-than-human worlds arises again in Hannah Rachel Cole's "A Hyphen Between Trees: Derek Walcott's Canoes." The canoe is a hyphen—both connector and divider—between trees and generations whose materiality, as well as symbolism, brings to the surface painful Caribbean ecological and colonial histories. Drawing on Amerindian, African, and European traditions, Walcott's poetry foregrounds the canoe as a mediator in historical trauma, from Indigenous resilience to the shadow of the Middle Passage. The making of these vessels out of Saint Lucian forest species—such as the *gommier* (*Dacryodes excelsa*) and red cedar (*Cedrela odorata*)—prompts ambivalent feelings of sacrifice, connection, and loss. The vanished but imagined conversations among trees in the forest echo the disappearance of Amerindian cultures and languages. But the canoe is also a vessel that, through the operations of poetic language, travels in time as well as space to actualize the reparative efforts of memory and commemoration. Cole's chapter evokes the afterlife of plants not as mere vehicles for human needs but as carriers of intermingled environmental and human histories.

Some of the residual power of Walcott's canoes derives from their material connection—as former revered trees—to Indigenous cosmologies, a subtle reminder of the spiritual significance of plants to human cultures. The last section on the Sacred explores past instantiations and future implications of this profound connection between plants and human spirituality for the plant humanities. In "Planting Buddhism in Medieval China," Natasha Heller examines the role of plants in the transplantation of Buddhism to medieval China, especially as evidenced in Yang Xuanzhi's *A Record of Buddhist Gardens in Luoyang*. If "church planting" in Protestant Christianity figured the establishment of new congregations, Heller invites us to consider the spread of Buddhism in China as "enforested." She points to the centrality of trees to the Buddha's life: he was born under a sāla tree (*Shorea robusta*), awakened under a bodhi tree (*Ficus religiosa*), and died between sāla trees. As Buddhism entered China, monasteries were referred to as "dense forests" (*conglin*) as well as gardens, and their trees became a primary form of Buddhist placemaking, adapted to Chinese climates and aesthetics. *A Record of Buddhist Gardens in Luoyang* mentions both individuated sacred trees, such as the unusual five-layered mulberry at Yuanhui Temple, and imported species, such as grapes and pomegranates that became objects of pilgrimage, wonder, and devotion. Everyday trees, such as pine, cypress, juniper, and bamboo, also shaped monastic experience. Plants thus became cocreators of Buddhist religious environments that fostered both spiritual contemplation and material hospitality and nourishment. In Heller's telling, the transplantation of Buddhism to China was also a process of botanical integration, where plants and gardens were not mere backdrops but active participants in the creation of sacred spaces and experiences.

Such deep and long-standing spiritual relationships between humans and plants are vital and fragile, even endangered. The last chapter by Gary Paul Nabhan, "Plant Humanities Join Ethnobotany to Reduce Cultural Parallax and More Fully Protect Sacred Plants?" makes a powerful case for the value of plants through the lens of the sacred. Nabhan focuses on the iconic saguaro cactus (*Carnegiea gigantea*) of the Sonoran Desert as a sacred plant of the O'odham peoples, by whom it is revered as a sentient being embedded in creation stories, spiritual practices, and seasonal ceremonies. The saguaro fruit, fermented into wine, serves as a sacramental element of rain-bringing rituals. Yet despite historic acknowledgment of the O'odham peoples' special relationship with the cactus, the devastation wrought by border wall construction in 2017–2021 endangered both saguaros and Indigenous access to them. In the wake of this destruction, Nabhan describes the struggle of O'odham leaders and communities for recognition of their legal right and spiritual relationship to the plant. The case of the saguaro is indicative of the current threat to sacred plants and their role in spiritual traditions across the globe. The Sacred Plant Biocultural Recovery Initiative, launched in 2024, endeavors to bring together Indigenous, interfaith, and academic leaders to record and communicate the value of sacred plants not just for plant conservation but for the persistence of these traditions. Plant humanities scholars, Nabhan urges, have a vital role to play in the urgent, cross-cultural effort to convey and protect the deep spiritual and cultural value of plants.

Notes

1 I am grateful to W. John Kress for a long-standing dialogue that has helped shape this introduction, my coeditor Anatole Tchikine for many illuminating conversations, our editor Sara Taylor for her extraordinary support and discernment, and two anonymous reviewers for their helpful comments. All errors are, of course, entirely my own.

2 Alexandre Antonelli, "State of the World's Plants and Fungi," *Kew Royal Botanic Gardens*, 2020, https://www.kew.org/science/state-of-the-worlds-plants-and-fungi; and Eimear Nic Lughadha, Steven P. Bachman, Tarciso C. C. Leão, et al., "Extinction Risk and Threats to Plants and Fungi," *Plants, People, Planet* 2, no. 5 (2020): 389–408, https://doi.org/10.1002/ppp3.10146.

3 Rob Nixon, *Slow Violence and the Environmentalism of the Poor* (Cambridge, Mass.: Harvard University Press, 2013); and Ashlee Cunsolo and Karen Landman, eds., *Mourning Nature: Hope at the Heart of Ecological Loss and Grief* (Montreal: McGill-Queen's University Press, 2017).

4 Michael Pollan, *The Botany of Desire: A Plant's Eye View of the World* (New York: Random House, 2001).

5 The literature on botany and empire is voluminous. See Londa Schiebinger, *Plants and Empire: Colonial Bioprospecting in the Atlantic World* (Cambridge, Mass.: Harvard University Press, 2004); Emma C. Spary, *Utopia's Garden: French Natural History from Old Regime to Revolution* (Chicago: University of Chicago Press, 2000); Lucile H. Brockway, *Science and Colonial Expansion: The Role of the British Royal Botanic Gardens* (New Haven: Yale University Press, 1979); Richard Harper Drayton, *Nature's Government: Science, Imperial Britain, and the "Improvement" of the World* (New Haven: Yale University Press, 2000); Kapil Raj, *Relocating Modern Science: Circulation and the Construction of Knowledge in South Asia and Europe, 1650–1900* (London: Palgrave Macmillan, 2007); Mary Louise Pratt, *Imperial Eyes: Travel Writing and Transculturation* (New York: Routledge, 1992); Londa Schiebinger and Claudia Swan, eds., *Colonial Botany: Science, Commerce, and Politics in the Early Modern World* (Philadelphia: University of Pennsylvania Press, 2005); David Philip Miller and Peter Hanns Reill, eds., *Visions of Empire: Voyages, Botany, and Representations of Nature* (Cambridge: Cambridge University Press, 1996); and Yota Batsaki, Sarah Burke Cahalan, and Anatole Tchikine, eds., *The Botany of Empire in the Long Eighteenth Century* (Washington, D.C.: Dumbarton Oaks Research Library and Collection, 2016).

6 Staffan Müller-Wille and Isabelle Charmantier, "Natural History and Information Overload: The Case of Linnaeus," *Studies in History and Philosophy of Science: Part C, Studies in History and Philosophy of Biological and Biomedical Sciences* 43, no. 1 (2012): 4–15.

7 Alfred W. Crosby, *The Columbian Exchange: Biological and Cultural Consequences of 1492* (Westport, Conn.: Greenwood Press, 1972); Alfred W. Crosby, *Ecological Imperialism: The Biological Expansion of Europe, 900–1900* (Cambridge: Cambridge University Press, 1986); Judith A. Carney and Richard N. Rosomoff, *In the Shadow of Slavery: Africa's Botanical Legacy in the Atlantic World* (Berkeley: University of California Press, 2009); and Judith A. Carney, *Black Rice: The African Origins of Rice Cultivation in the Americas* (Cambridge, Mass.: Harvard University Press, 2001).

8 Schiebinger and Swan, *Colonial Botany*.

9 Brian W. Ogilvie, *The Science of Describing: Natural History in Renaissance Europe* (Chicago: University of Chicago Press, 2006).

10 Brockway, *Science and Colonial Expansion*.

11 On the Plantationocene, see Donna Haraway, "Anthropocene, Capitalocene, Plantationocene, Chthulucene: Making Kin," *Environmental Humanities* 6, no. 1 (2015): 159–65. For a more inclusive view of the plantation in historical and geographical context, see Katherine McKittrick, "Plantation Futures," *Small Axe: A Journal of Criticism* 17, no. 3 (2013): 1–15; and Mythri Jegathesan, "Black Feminist Plots Before the Plantationocene and Anthropology's 'Regional Closets,'" *Feminist Anthropology* 2, no. 1 (2021): 78–93. A great introduction to plantation studies and relevant scholarly literature is Sophie Chao, "Plantation," *Environmental Humanities* 14, no. 2 (2022): 361–66.

12 Val Plumwood, *Feminism and the Mastery of Nature* (London: Routledge, 1993); Val Plumwood, *Environmental Culture: The Ecological Crisis of Reason* (London: Routledge, 2002); Carolyn Merchant, *The Death of Nature: Women, Ecology, and the Scientific Revolution* (San Francisco: Harper & Row, 1980); and Carolyn Merchant, *Reinventing Eden: The Fate of Nature in Western Culture* (New York: Routledge, 2003).

13 On the ontological devaluation of plants, see Michael Marder, *Plant-Thinking: A Philosophy of Vegetal Life* (New York: Columbia University Press, 2013); Matthew Hall, *Plants as Persons: A Philosophical Botany* (Albany: State University of New York Press, 2011); and Jeffrey Nealon, *Plant Theory: Biopower and Vegetable Life* (Stanford: Stanford University Press, 2016). Other important entry points to the growing conversation in critical plant studies include Prue Gibson, *The Green Thread: Dialogues with the Vegetal World* (Lanham, Md.: Lexington Books, 2015); Prue Gibson and Margaret Somerville, eds., *Vegetal Entanglements: Vegetal Ontologies in Plant Science, Philosophy, Literature, Art, and Ecology* (Adelaide: punctum books, 2022); and Giovanni Aloi and Michael Marder, eds., *Vegetal Entwinements in Philosophy and Art* (Cambridge: Cambridge Scholars Publishing, 2021).

14 Amitav Ghosh, *The Nutmeg's Curse: Parables for a Planet in Crisis* (Chicago: University of Chicago Press, 2021); and Anna Lowenhaupt Tsing, *The Mushroom at the End of the World: On the Possibility of Life in Capitalist Ruins* (Princeton: Princeton University Press, 2015).

15 Anna Lowenhaupt Tsing, Heather Swanson, Elaine Gan, and Nils Bubandt, "Patchy Anthropocene: Landscape Structure, Multispecies History, and the Retooling of Anthropology," *Current Anthropology* 60, no. S20 (2019): S186–97; Anna Lowenhaupt Tsing, Heather Swanson, Elaine Gan, and Nils Bubandt, eds., *Arts of Living on a Damaged Planet:*

Ghosts and Monsters of the Anthropocene (Minneapolis: University of Minnesota Press, 2017); Robin Wall Kimmerer, *Braiding Sweetgrass: Indigenous Wisdom, Scientific Knowledge, and the Teachings of Plants* (Minneapolis: Milkweed Editions, 2013); Robin Wall Kimmerer, *Gathering Moss: A Natural and Cultural History of Mosses* (Corvallis: Oregon State University Press, 2003); and Robin Wall Kimmerer, "The Serviceberry: An Economy of Abundance," *Orion* 39, no. 1 (2020): 12–26.

16 Ghosh, *Nutmeg's Curse*, 76, 204.

17 Yota Batsaki, "Introducing the Plant Humanities Lab," *Harvard Library Bulletin*, 2021.

18 Mark Nesbitt, "Biocultural Collections: Needs, Ethics and Interdisciplinary Collaborations," *Ethnobotany Research & Applications* 12 (2014): 1–11; Caroline Cornish, *Curating Science in an Age of Empire: Kew's Museum of Economic Botany* (London: Pickering & Chatto, 2013); Caroline Cornish, Mark Nesbitt, and Paul A. Rees, eds., *Curating Biocultural Collections: A Handbook* (London: Kew Publishing, 2014); Felix Driver and Brenda S. A. Morton, "Mobility and the Production of Ethnographic Knowledge: Inventing the British Empire's Plant Hunters," *Journal of Historical Geography* 49 (2015): 63–74; and Felix Driver, Mark Nesbitt, and Caroline Cornish, eds., *Mobile Museums: Collections in Circulation* (London: UCL Press, 2021).

19 Pamela Soltis, "Digitization of Herbaria Enables Novel Research," *American Journal of Botany* 104, no. 9 (2017): 1281–84; and Carly R. Cowell, Pippin M. L. Anderson, Wendy A. Annecke, and Richard Ladle, "Historic Herbarium Specimens as Biocultural Assets: An Examination of Herbarium Specimens and Their In Situ Plant Communities of the Agulhas National Park, South Africa," *People and Nature* 2, no. 2 (2020): 483–94. For two more encompassing accounts of the herbarium, see Barbara M. Thiers, *Herbarium: The Quest to Preserve and Classify the World's Plants* (Portland, Ore.: Timber Press, 2020); and Maura C. Flannery, *In the Herbarium: The Hidden World of Collecting and Preserving Plants* (New Haven: Yale University Press, 2023).

20 J. Mason Heberling and Bonnie L. Isaac, "Herbarium Specimens as Exaptations: New Uses for Old Collections," *American Journal of Botany* 104, no. 7 (2017): 963–65.

21 Ghillean T. Prance and Mark Nesbitt, eds., *The Cultural History of Plants* (New York: Routledge, 2005), 3; and Robert Sussman, D. Tab Rasmussen, and Peter H. Raven, "Rethinking Primate Origins Again," *American Journal of Primatology* 75, no. 2 (2013): 95–106, https://doi.org/10.1002/ajp.22096.

22 James H. Wandersee and Elisabeth E. Schussler, "Preventing Plant Blindness," *The American Biology Teacher* 61, no. 2 (1999): 82–86; James H. Wandersee and Elisabeth E. Schussler, "Toward a Theory of Plant Blindness," *Plant Science Bulletin* 47, no. 1 (2001): 2–9; Kathryn M. Parsley, "Plant Awareness Disparity: A Case for Renaming Plant Blindness," *Plants, People, Planet* 2, no. 6 (2020): 598–601; and Elinor M. Breman, Stephanie Atkinson, and Sara Waldron, "Plant Blindness and the Implications for Plant Conservation," *Plants, People, Planet* 4, no. 2 (2022): 227–31.

23 See, for example, Harold C. Conklin, *The Ethnobotany of the Hanunóo: A Study in the Meaning of Primitive Plant Classification* (New York: Viking Fund, 1954).

24 Prance and Nesbitt, *Cultural History of Plants*, 11.

25 Pollan, *Botany of Desire*.

26 See, for example, Cassandra Leah Quave, *The Plant Hunter: A Scientist's Quest for Nature's Next Medicines* (New York: Viking, 2021).

27 "WHO Establishes the Global Centre for Traditional Medicine in India," news release, March 25, 2022, https://www.who.int/news/item/25-03-2022-who-establishes-the-global-centre-for-traditional-medicine-in-india.

28 Prance and Nesbitt, *Cultural History of Plants*, 173.

29 Ioannis Fappas, "The Oil of Gladness: The Ancient Tradition of the Manufacture of Aromatic Oils in the Aegean and the Eastern Mediterranean," in *Cutting-Edge Technologies in Ancient Greece*, ed. Marina Panagiotaki, Ilias Tomazos, and Fotios Papadimitrakopoulos (Oxbow: Oxbow Books, 2020), 113.

30 Julia Fine, "Maize: Sacred Plant, Global Commodity," *Plant Humanities Lab*, https://lab.plant-humanities.org/maize/; and Liza Grandia, "Sacred Maize Against a Legal Maze: The Diversity of Resistance to Guatemala's 'Monsanto Law,'" *Journal for the Study of Religion, Nature and Culture* 11, no. 1 (2017): 56–85. I am grateful to Frauke Sachse for enlightening conversations on maize.

31 Fine, "Maize."

32 Harris A. Lewin, Gene E. Robinson, W. John Kress, et al., "Earth BioGenome Project: Sequencing Life for the Future of Life," *Proceedings of the National Academy of Sciences* 115, no. 17 (2018): 4325–33.

33 See W. John Kress and Yota Batsaki, "Plant of the Month: Heliconia," *JSTOR Daily*, October 21, 2020, https://daily.jstor.org/plant-of-the-month-heliconia/. For the literature on plant cognition, see Zoë Schlanger, *The Light Eaters: How the Unseen World of Plant Intelligence Offers a New Understanding of Life on Earth* (New York: Harper, 2024); Monica Gagliano, *Thus Spoke the Plant: A Remarkable Journey of Groundbreaking Scientific Discoveries and Personal Encounters with Plants* (Berkeley, Calif.: North Atlantic Books, 2018); Stefano Mancuso and Alessandra Viola, *Brilliant Green: The Surprising History and Science of Plant Intelligence* (Washington, D.C.: Island Press, 2015); Daniel Chamovitz, *What a Plant Knows: A Field Guide to the Senses* (New York: Scientific American / Farrar, Straus and Giroux, 2012); Paco Calvo and Natalie Lawrence, *Planta Sapiens: The New Science of Plant Intelligence* (New York: W. W. Norton, 2023); and Anthony Trewavas, *Plant Behaviour and Intelligence* (Oxford: Oxford University Press, 2014).

34 For a visual record, see W. John Kress and Shirley Sherwood, *The Art of Plant Evolution* (Richmond: Kew Publishing, 2009).

35 Lyanda Lynn Haupt, *Rooted: Life at the Crossroads of Science, Nature, and Spirit* (New York: Little, Brown Spark, 2021), 154–55.

36 On the cultural and environmental history of kudzu in the American South, see Derek H. Alderman, "Channing Cope and the Making of a Miracle Vine," *Geographical*

Review 94, no. 2 (2004): 157–77; Derek H. Alderman, "When an Exotic Becomes Native: Taming, Naming, and Kudzu as Regional Symbolic Capital," *Southeastern Geographer* 55, no. 1 (2015): 32–56; and J. Winberry and D. M. Jones, "Rise and Decline of the 'Miracle Vine': Kudzu in the Southern Landscape," *Southeastern Geographer* 13, no. 2 (1973): 61–70.

37 Natasha Myers, "From Edenic Apocalypse to Gardens Against Eden: Plants and People In and After the Anthropocene," in *Infrastructure, Environment, and Life in the Anthropocene*, ed. Kregg Hetherington (Durham: Duke University Press, 2019), 124; and Bettina Stoetzer, *Ruderal City: Ecologies of Migration, Race, and Urban Nature in Berlin* (Durham: Duke University Press, 2022).

38 Gert Gröning and Joachim Wolschke-Buhlman, "The Native Plant Enthusiasm: Ecological Panacea or Xenophobia?," *Landscape Research* 28, no. 1 (2003): 75–88, https://doi.org/10.1080/01426390306536; and Banu Subramaniam, "The Aliens Have Landed: Reflections on the Rhetoric of Biological Invasions," *Meridians* 2, no. 1 (2001): 26–40.

39 Benedict Anderson, *Imagined Communities: Reflections on the Origin and Spread of Nationalism* (London: Verso, 2016).

40 Jacques Derrida, *Archive Fever: A Freudian Impression*, trans. Eric Prenowitz (Chicago: University of Chicago Press, 1996).

Mobilities

2

AFRICAN FOODS IN DIASPORA

Sustenance, Worldmaking, and Neotropical Legacies

JUDITH A. CARNEY AND RICHARD N. ROSOMOFF

The story of African food plants in the Americas is inextricably entangled with the history of New World slavery. It is a story of sustenance, survival, and freedom that emerges from the shattered life worlds of the enslaved. From the beginnings of plantation societies in the sixteenth century, seeds and root crops from tropical Africa stealthily infused a subaltern food system that slaves created to sustain themselves amid the brutal labor regime to which they were condemned. By interplanting these cultivars with others of Amerindian and European origin, the enslaved shaped a distinctive and robust food system that sharply contrasted with the plantation monocultures of the era.[1]

Today, this heritage food system still endures in many Afro-descendant rural communities. It is imperiled by the renewed embrace of monocultures and the rapid expansion of soy and oil palm plantations across the New World tropics. Indeed, plantations past and present have so influenced land-use futures that the modern era is often labeled the "Plantationocene" instead of the "Anthropocene" to underscore the ecosystem destruction, biodiversity loss, and racial tensions that continue to shadow industrial agriculture. Chemically dependent large-scale monocultures are devastating tropical biomes, where the globe's richest biodiversity is concentrated, and accelerating species extinctions. Their spread is sweeping aside alternative, time-tested farming systems whose rich repositories of agroecological knowledge and practices augur more flexible, resilient, and equitable food futures.[2]

In the New World tropics, many of these endangered land-use systems have histories that are both compelling and relevant. In the formative period of plantation slavery, food insecurity was a brutal fact of daily existence. Hunger, malnutrition, and overwork took a terrible toll among the chattelized workforce. Out of necessity, enslaved Africans began cultivating subsistence foods in plantation spaces unsuited to the production of the export

FIGURE 2.1
Afro-descendant food garden, Jamaica, 1978. Photograph by Judith A. Carney.

commodity, whether sugar, coffee, indigo, cotton, or rice. They sowed these spaces with a diverse array of grains, tubers, vegetables, and fruit trees, creating a multistory canopy structure mimicking the luxuriant vegetative architecture of a tropical forest that naturally protected crops from insect predation, heavy downpours, and soil erosion. The apparent chaos of these food forests often confused and confounded European observers, for they weren't planted with the geometric rigor of commodity production, making it "difficult to distinguish garden from forest."[3] This planting strategy is technically known as a polyculture, a medley of food and medicinal plants cultivated together in a mixed or intercropped system. After abolition, free Blacks transitioned this subaltern food regime to an agrobiodiverse peasant farming system since practiced by generations of Afro-descendant people (Figure 2.1).[4]

A defining feature of this alternative food system is the number of plants that arrived from Africa. Table 2.1 lists more than two dozen African plants that were cultivated in slave food plots. Among some notable introductions are millet, sorghum, yams, okra, oil palm, black-eyed and pigeon peas, watermelon, hibiscus, and rice. The African contributions to this distinctive New World polyculture raise a provocative question: How did these Old World tropical plants arrive in the New World and become key components of the subaltern food system developed by the enslaved?

The Columbian Exchange, Plant Mobilities, and African Agency

The decades following 1492 were momentous on several levels. As Europeans arrived in the Americas and began imperial wars of conquest and settlement, their overseas expansion led to an unprecedented exchange of plants, animals, and microbes across several continents. Historian Alfred W. Crosby famously called this movement of species—deliberate and accidental, seen and unseen—the Columbian Exchange. Crosby was interested not only in the novel species that naturalists brought back to Europe but also in the plants and animals that European migrants introduced to the Americas. His research recognized that the agents of biological and botanical translocation were not just prominent men of science but ordinary people as well.[5]

However, a less heralded, but no less important, intercontinental transfer of species occurred at the same time. In this instance, the agents of plant establishment were African; the migrants were enslaved; and the plant transfers involved tropical species from Africa. These African introductions were unlike any other discussed in standard Columbian Exchange accounts, for they unfolded as a direct consequence of the Atlantic slave trade, which lasted for 350 years and forcibly migrated some 12.5 million Africans to the Americas on more than thirty-five thousand slave voyages.[6]

TABLE 2.1

African food plants established in the plantation societies of the New World.

COMMON NAME	SCIENTIFIC NAME	ORIGIN
CEREALS		
Millet	*Pennisetum glaucum*	
Sorghum	*Sorghum bicolor*	
Rice*	*Oryza glaberrima* *Oryza sativa*	 Asian
TUBERS		
Yams	*Dioscorea cayenensis*	
Plantain/Banana	*Musa* spp.	Asian
Taro/Eddo	*Colocasia esculenta*	Asian
LEGUMES		
Black-eyed pea/Cowpea/ Calavance	*Vigna unguiculata*	
Pigeon pea/Angola pea/Congo pea/*Guandú*	*Cajanus cajan*	Asian
Bambara groundnut/*Voandzeia*	*Vigna subterranea*	
Lablab bean/Hyacinth bean/Bonavist bean	*Lablab purpureus*	
OIL, PLANTS, AND FRUITS		
Sesame seeds/Benne seeds	*Sesamum radiatum*	Asian
Castor bean	*Ricinus communis*	
Oil palm	*Elaeis guineensis*	
Watermelon	*Citrullus lanatus*	
Muskmelon	*Cucumis melo*	
Ackee	*Blighia sapida*	
BEVERAGES		
Coffee	*Coffea* spp.	
Tamarind	*Tamarindus indica*	Asian
Hibiscus/Sorrel	*Hibiscus sabdariffa*	
Kola nut	*Cola* spp.	
VEGETABLES AND SPICES		
Okra	*Abelmoschus esculentus*	
Spinach/Callaloo/Jute mallow	*Amaranthus* spp./*Corchorus olitorius*	
Melegueta pepper/Guinea grains	*Aframomum melegueta*	
Guinea pepper	*Xylopia aethiopica*	
Guinea squash	*Solanum aethiopicum*	

*African rice (*Oryza glaberrima*) is not the same as Asian rice (*Oryza sativa*). It is a separate species that was independently domesticated in West Africa more than three thousand years ago but not confirmed as a separate species until the mid-twentieth century. Plants of Asian origin were introduced to Africa during prehistory and adopted into African food systems. Judith A. Carney and Richard N. Rosomoff, *In the Shadow of Slavery: Africa's Botanical Legacy in the Atlantic World* (Berkeley: University of California Press, 2009), 136–37.

One of the striking features of the history of the formative plantation period is the number of European accounts that credit the enslaved with the introduction of specific foods, all previously grown in Africa. Naturalists and visitors to plantation societies claimed that slaves introduced at least two dozen plants, including yams, sorghum, millet, sesame, black-eyed and pigeon peas, kola nut, and okra. Reading the commentaries penned by diverse historical figures such as Hans Sloane, Willem Piso, Georg Marcgrave, Thomas Jefferson, Mark Catesby, and other naturalists who made these claims begs the question: How is it possible that slaves introduced crops to the Americas?[7] After all, Africans were brought to plantation societies in chains and without personal possessions. Crediting enslaved Africans with plant introductions attributes to them an agency that is entirely at odds with longstanding assumptions. So why would these prominent historical witnesses claim otherwise?[8]

Despite the appalling human dislocations that slavery visited on the continent, Africa often produced food for sale to slave ships. This is known from ship manifests as well as from drawings and logbooks of ship captains.[9] Slave ships carried some food supplies from Europe, but captains also relied on food grown in Africa—including sixteenth-century Amerindian introductions, such as maize and peanuts—to provision enslaved captives across the Middle Passage. The transatlantic journey could last six weeks or more, so sufficient food stores were vital. Chief among the dietary staples routinely purchased were African rice, yams, sorghum, and millet.[10] Certainly, most slave voyages did not end with leftover provisions. But those that did presented slaves with the opportunity to access seeds and rootstock of familiar crops and to reestablish them in new lands. These staples soon appeared in the food gardens of the enslaved. Thus, as Africans arrived in the Americas, so did African foods.

How did these African foods find a footing in New World plantation societies? The written record is, for the most part, silent. However, oral histories still held by communities descended from runaway slaves, known as Maroons, offer an intriguing explanation. These stories, which have been passed down from one generation to the next, attribute the arrival of a key African food staple, rice, to the actions of enslaved women who smuggled seeds off the slave ship to reinstate a familiar dietary staple. In this way, African crops entered the food

FIGURE 2.2

Slave Ship *Fredensborg II*, 1788, detail. Reproduced from *Slavery Images*, https://www.slaveryimages.org/database/image-result.php?objectid=799.

FIGURE 2.3

Kumona Baba hand-milling rice, The Gambia, 1984. Photograph by Judith A. Carney.

regimen slaves created for their own sustenance. As with many oral histories, there is surprising convergence between folklore and the archive.

We know that captains of slavers arriving along the western African coast purchased rice, sorghum, and millet in both milled and unmilled forms. The distinction between milled and unmilled is quite important because an unmilled grain is also a seed that can be germinated. Once at sea, any grain that had been purchased unmilled—that is, still in its husk—required removing the indigestible hulls to ready it for human consumption. Slave ships would stock milling devices appropriate for their food stores: grinding stones or iron rollers for maize, the mortar and pestle for unhusked African cereals.[11]

During the Middle Passage, enslaved women were often made to work preparing meals and, to this end, ship captains aimed to board a complement of "Women to dress the Victuals."[12] Preparation typically took place above deck near the ship's stern, behind the barricade that kept male and female slaves apart. Females given the arduous job of hand-milling unhusked grain would have been tasked with pounding the cereal in a wooden mortar to remove the hulls, a skill indelibly associated with women's work in Africa. The journal of the slave ship *Mary* mentioned females cleaning rice on its outbound 1795 voyage from the Indigenous West African rice region.[13] Two decades earlier, British naturalist Henry Smeathman, returning from Sierra Leone on a slaver, wrote poignantly of the captives' shipboard misery while observing enslaved females being pressed into meal service: "Alas! What a scene of misery and distress is a full slaved ship in the rains. The clanking of chains, the groans of the sick and the stench of the whole is scarce supportable . . . two or three slaves thrown overboard every other day dying of fever, flux, measles, worms all together. All the day the chains rattling or the sound of the armourer rivetting some poor devil just arrived in galling heavy irons. The women slaves in one part beating rice in mortars to cleanse it for cooking."[14] In 1796, Dr. George Pinckard also witnessed slaves milling rice during a visit aboard an American slave ship that had stopped in Barbados en route to Savannah, Georgia: "We saw several of them employed in beating the red husks off the rice, which was done by pounding the grain in wooden mortars, with wooden pestles, sufficiently long to allow them to stand upright while beating in mortars placed at their feet."[15]

This work is portrayed in an eighteenth-century painting of the Danish slave ship *Fredensborg* (Figure 2.2). Close examination of the quarterdeck shows the outlines of two women lifting wooden pestles. The detail clearly shows them milling grain in the traditional African way, striking it with pestles in a wooden mortar to separate each grain from its husk (Figure 2.3). In this painting, it is not known if the women are hand-milling sorghum, millet, or rice, but it is certain that none of these African crops were present in the Americas prior to the arrival of Europeans and Africans. The proximity of enslaved women to the slave ship's food stores, their role in meal preparation, and their access to seed grain resonates with Maroon oral histories and their agential female protagonists.

In isolated Maroon hamlets from Suriname to Brazil, rice continues to be esteemed as a food and ceremonial offering. Despite the vast distances that separate them, remarkably congruent stories are told about how these communities came to grow rice. The accounts recognize rice as a crop from Africa and attribute its introduction to an enslaved female forebear who hid seeds in her hair as she disembarked the slave ship. Other narratives elaborate on the onset of rice cultivation by highlighting a female plantation runaway who similarly conceals the precious grains and carries them to fugitive communities. Many of the rice varieties still grown by Suriname's Maroons are known by the names of these female ancestors.[16]

Across northeastern South America, Maroon oral histories reclaim rice as an African origin story, making explicit its connection to slave ships, the Middle Passage, African initiative, dietary preferences, and diasporic identity. In these societies, as in much of West Africa, women have always had a central role in rice farming.[17] By valorizing enslaved African women as the agents of rice transfers and cultivation, the Maroon narratives collectively illuminate a layer of memory and historical consciousness that decenters standard Eurocentric accounts of crop transfers in which African agency and initiative are excluded.

During the eighteenth century, colonial militias were increasingly mobilized to capture growing numbers of escaped slaves and to clear frontier hinterlands of fugitive enclaves. Expedition reports occasionally mention the food fields surrounding the communities they subdued and the significance of rice as a dietary staple. Moravian missionaries proselytizing among Suriname's Saramaka Maroons in the 1770s echoed these reports; they also found rice to be widely cultivated.[18]

Following scientific consensus that African rice (*Oryza glaberrima*) is, in fact, a separate and independently domesticated species, scholars in the twenty-first century began to investigate Maroon claims about its African introduction. An initial botanical expedition to Suriname in 2008 collected *glaberrima* rice from a community's farm plot, where it was being cultivated by women as an offering in ancestor rituals. Genomic studies trace this New World *glaberrima* specimen and others subsequently found in Maroon fields back to the West African rice region, with evidence that African rice was introduced to Suriname during the period of plantation slavery.[19]

Food Gardens Amid Plantation Monocultures

Historic illustrations of Caribbean plantations commissioned to romanticize (and sanitize) the estate often portray slave food forests in plain sight, even though to the viewer they appear as little more than scattered arboreal decorations (Figure 2.4).[20] Nevertheless, these artworks offer visual clues to the existence of a subaltern food regime within the interstices of the plantation landscape. As to what might lie beneath these picturesque canopies, a nineteenth-century painting from Brazil provides context (Figure 2.5). It shows a slave dooryard garden and the variety of plants the enslaved cultivated around their humble dwellings. Tropical cultivars from Africa and the Americas abound, illustrating the creolization of the Old and New World plant heritages that slaves drew upon to develop their

FIGURE 2.4
Montpelier Estate, St. James's, from James Hakewill, *A Picturesque Tour of the Island of Jamaica, from Drawings Made in the Years 1820 and 1821* (London: Hurst and Robinson, 1825), pl. 19. Reproduced from *Slavery Images*, https://www.slaveryimages.org/database/image-result.php?objectid=472.

FIGURE 2.5
Slave Houses, Brazil, 1830s, from Johann Moritz Rugendas, *Voyage pittoresque dans le Brésil* (Paris: Engelmann, 1835). Note the papaya and plantains surrounding the dwelling, the castor plant to the left of the two walking figures, and the pineapples growing at their feet. Reproduced from *Slavery Images*, https://www.slaveryimages.org/database/image-result.php?objectid=483.

distinctive planting system. The eighteenth-century British planter Bryan Edwards described similar forest gardens in Jamaica, observing that slave quarters were "always intermingled with fruit-trees."[21]

Slave dooryard gardens also sheltered cultivars requiring additional care and inputs, typically vegetables, herbs, spices, and medicinals. Some slaves kept small food animals whose manure improved the garden's productivity. Pinckard described these kitchen plots during his visit to Barbados in 1796: "On these small patches of garden it is common for the slaves to plant fruits and vegetables, and to raise stock. Some of them keep a pig, some a goat, some guinea fowls [an introduced species native to Africa], ducks, chickens, pigeons, or the like."[22]

Many New World plantation societies also allowed slaves a separate plot of land to farm and a free day (often Sunday) each week to tend it. Though this practice effectively shifted the burden of subsistence away from plantation owners (thereby saving them money) and onto the enslaved themselves, these modest grants of labor autonomy enabled slaves to expand food cultivation beyond slaveholder-organized provision grounds and the kitchen gardens of the enslaved (Figure 2.5) to plots on marginal land or plantation peripheries that they could independently manage (Figure 2.6).[23] Independent plots could be on erosive hillslopes, in areas of depleted or stony soils, or in swamplands. They were often remotely sited and planted to crops that did not require daily visits or intensive labor, such as fruit trees, yams, taro, and plantains. In such pockets of slave subsistence cultivation—dooryard gardens, provision grounds, and independent plots—the enslaved innovated and affirmed a radically different food order from that of the monoculture on which they labored. In contrast, slaves' food-first system emphasized polycultures and agrobiodiversity.

In pre-revolutionary eighteenth-century Saint Domingue (Haiti), one French official referred to slave-managed food plots as "petite Guinées"—Guinea was the name by which Africa was known at the time—since the enslaved themselves decided what to grow and how to grow it. One missionary to the Danish West Indies in 1768 recorded slaves growing in their gardens more than eighteen different plants, many now known to be African

FIGURE 2.6

View of a Sugar Plantation, French West Indies, 1762, from Denis Diderot, *Encyclopédie, ou Dictionnaire raisonné des sciences, des arts et métiers*, 17 vols. (Paris: Chez Briasson, 1762–1772), 1:pl. 1. The mountainous area at the upper left (nos. 13 and 14) corresponds to independent food plots that the enslaved established on remote hillslopes of a sugar plantation; these areas were often planted with tropical root crops and tubers. Reproduced from *Slavery Images*, https://www.slaveryimages.org/database/image-result.php?objectid=1201.

introductions, such as okra, sorghum, yams, and pigeon peas.[24]

It was in such cultivation sites that slaveholders and naturalists in the New World often had their first encounters with unfamiliar tropical crops whose African provenance we now recognize. With no names for these novel plants in European languages, they borrowed the words used by the slaves who grew them. In this way, some African vernacular crop names passed into New World colonial languages. For example, in many West African languages, the word for yam is *inhame*. The Portuguese adopted this word, which became *ñame* in Spanish and *yam* in English. In English, *okra*, *ackee*, and *bissy* (for kola nut) are loanwords from the Akan language of Ghana. The Carolina colony adopted the Senegambian Wolof word *benne* for sesame. Portuguese Brazil appropriated the Bantu words *dendê* for oil palm and *guandu* for pigeon pea. Even *banana*, a word of unspecified West African origin, represents another African linguistic borrowing.[25]

Independent Production as Liberatory Practice

For the enslaved, self-provisioning of daily food needs ultimately became a kind of liberatory practice that anticipated and informed ongoing and future struggles for freedom. Food scarcity and starvation had plagued the formative period of plantation development. As slaves pressed for the right (and the time) to grow their own food, their petitions were sometimes successful, especially when there was infertile land on which the export crop could not grow. Planters found the arrangement advantageous because it reduced operating costs, deepened slaves' ties to the land, and discouraged marronage.[26] While this put the enslaved entirely in charge of their own subsistence, it established an important convention—namely, time each week exempted from plantation work to tend their plots. This small fissure in an otherwise inflexible plantation labor regime also secured for the enslaved the right to keep what

FIGURE 2.7
Marketplace, Antigua, West Indies, 1806. Reproduced from *Slavery Images*, https://www.slaveryimages.org/database/image-result.php?objectid=168.

they produced and to realize the benefits from any surpluses they might sell, along with small animals, in weekly markets (Figure 2.7). Anthropologists Sidney W. Mintz and Douglas Hall observed that slaves' independent plots stimulated the regional food networks that supplied plantation societies with meat and fresh produce, which were otherwise imported or in limited supply.[27] These weekly markets allowed the enslaved to intermingle, develop wider social networks, exchange knowledge about the world beyond the plantation, and actuate strategies of resistance.

Slave food gardens were also important for providing the seeds and root cuttings that runaways took with them when they fled bondage. A key to self-liberation was the ability to self-provision, for no fugitive community, no matter how well isolated or defended, could survive without a sustainable food and water supply. Maroon settlements were commonly sited away from main river channels, often near a protective swamp that could be planted to rice and plantains, and in proximity to a reliable source of fresh water, such as a stream. Despite the constant threat of discovery and attack, some settlements were able to survive for generations, others up to and beyond abolition. But more often than not, they were vanquished, and the community structures and fields were razed to prevent reoccupation.[28]

Militia reports from the eighteenth century occasionally described the system of "escape agriculture" practiced by Maroons. After destroying a settlement in Suriname's rainforest interior, one expedition recorded fields planted to rice, yams, and plantains. In the same century, Moravian missionaries listed some three dozen plants that Saramaka Maroons cultivated in their food gardens, of which one-third belonged to the dietary traditions of Africa. Along with rice, plantains, and yams, they found the Bambara groundnut (*Vigna subterranea*), okra, pigeon pea, watermelon, taro, ginger, sesame, and eggplant—all plants that were then (as now) part of African foodways.[29] Punitive military campaigns in nineteenth-century Maranhão, Brazil, describe fugitives cultivating a similar array of crops. Among the food and medicinal plants mentioned are manioc, maize, sweet potatoes, pineapples, sugarcane, rice, sesame, millet, plantains, castor bean, and *diamba* (marijuana).[30] The last six crops were African introductions to colonial Maranhão. Together, these Old and New World tropical cultivars elaborated creolized plant ensembles that had diffused from plantation slave plots to the food fields of self-liberated communities.

Brazil abolished slavery in 1888; it was the last country in the Western Hemisphere to do so. After emancipation, this subaltern food system endured (as it did elsewhere in former plantation societies of the New World tropics), informing the farming strategies by which freed persons sought to escape new exploitative labor relations and reconstituted racial hierarchies. In this regard, slave food plots can be seen not only as generative sites where the enslaved enacted and performed autonomous activities but also as crucibles of liberatory practice in which the long struggle for freedom was activated.

Biocultural Refugia and Planetary Food Futures

In the post-emancipation period, the Afro-descendant system of tropical farming continued to be practiced by the new peasant class of freed slaves (Figure 2.8). It also continued to perplex visitors from temperate climes, just as it had before. Writing about Trinidad in 1869, three decades after emancipation, the English clergyman Charles Kingsley was struck by the lush anarchy of the food forest and its ability to sustain Black families, even though he considered their agricultural methods unsophisticated and inefficient. By his own admission, Kingsley found it difficult to distinguish individual crops and areas of cultivation from the surrounding habitat: "The 'provision grounds' of the Negroes were very interesting. I had longed to behold, alive and growing, fruits and plants which I had heard so often named, and seen so often figured, that I had expected to recognise many of them at first sight; and found, in nine cases out of ten, that I could not."[31]

It took nearly a century for the prevailing view of Afro-descendant polycultures as inchoate and irredeemable to be reconsidered. Geographers and anthropologists in the 1960s undertook systematic studies of peasant farms in the Caribbean. This followed mounting recognition of Indigenous peoples as experienced tropical farmers and stewards of agrobiodiversity at a time when chemically

FIGURE 2.8
Trinidadian mixed crop cultivation of cacao and banana, 1903. Reproduced from Brandon Head, *The Food of the Gods: A Popular Account of Cocoa* (London: R. Brimley Johnson, 1903).

dependent "green revolution" technology was upending traditional agriculture throughout the tropics.[32] The studies of Caribbean kitchen gardens and farms countered long-held declensionist views of Afro-descendant farmers as inept or, at best, desultory executors of food systems that Native peoples had perfected over millennia. The new research complemented work on Amerindian polycultures by documenting the agrobiodiversity present in this peasant farming system. A typical Afro-descendant farm in these Caribbean studies included as many as twenty to thirty different crops.[33]

More recent scholarship extends the conversation about Native peoples as crucial stewards of agrobiodiversity to other cultural groups whose histories of growing food emerged under entirely different socio-ecological circumstances.[34] In Brazil, this has stimulated research on Afro-descendant agricultural systems, especially the agrobiodiversity practiced and maintained by Maroon (*quilombo*) communities. A survey of one *quilombo* community's food gardens in Minas Gerais registered eighty-four species and landraces planted for subsistence. Figure 2.9 shows a similarly diverse crop assemblage on a *quilombo* farm in Maranhão.[35]

FIGURE 2.9

Francisco da Conceição cutting rice with a small knife in a *quilombo* food garden, Maranhão, Brazil, 1997. Photograph by Judith A. Carney.

The land-use histories of all these groups constitute biocultural heritages that were formed and accrued in situ, proven over time, and transmitted across generations to secure food availability and cultural resilience. What distinguishes this Afro-descendant food system from that of Amerindians is its fusion of cultigens and practices that exemplify the agrarian achievements of tropical peoples on both sides of the Atlantic. The Afro-descendant portfolio of place-based agroecological knowledge, shaped and maintained during centuries of hardship and experimentation, has endured under oppressive regimes of slavery and post-slavery histories of Black disenfranchisement.

Writing about heritage landscapes in Europe, Stephan Barthel, Carole Crumley, and Uno Svedin propose the concept of *biocultural refugia*, defined as "places that not only shelter species, but also carry knowledge and experiences about practical management of biodiversity and ecosystem services."[36] Biocultural refugia are the product of a coevolving human-environmental relationship that over time facilitates food security in specific places by nurturing biodiversity as an agricultural strategy. While the concept seeks to protect esteemed cultural landscapes of smallholders in Europe, it is also applicable to many Afro-descendant heritage farmlands in the New World tropics. This food system is notable for having initially developed on marginal or depleted plantation soils or under the adversities of escape agriculture. It is infused with knowledge and experiences shaped by social memories of slavery, food insecurity, and struggles for freedom. These resilient qualities enhance the capacity of Afro-descendant biocultural refugia to absorb socio-ecological shocks and to adapt without losing fundamental functions.

Today, Afro-descendant biocultural refugia tend to exist in highly contested rural areas where powerful political and economic forces threaten their survival and the agrobiodiversity they safeguard. Maroon communities have, for the most part, been unable to secure legal title to land they've occupied for centuries. Even when land titles have been confirmed, they nevertheless remain vulnerable to land grabs for dam projects, industrial mining operations, and large-scale agro-industrial monocultures. In Brazil, nowhere do these tensions play out more prominently than in the ongoing struggle by *quilombolas* who petition for collective rights to their ancestral farmland. Brazil officially recognized this right in its 1988 constitution—exactly one hundred years after abolition—but full legal investiture has proceeded at a glacial pace. Of the 3,563 *quilombos* registered nationally, the federal government has accorded limited recognition to only 494, or 7 percent of the total.[37] In this respect, advancing the biocultural refugia concept on behalf of *quilombo* farmscapes not only protects a remarkable legacy food system but also acknowledges their heritage land rights.

Conclusion

The cultural history of plants and their entangled links with human societies asks us to be attentive to the places and spaces where the disenfranchised have, or are, able to effect challenges to dominant food orders. Moreover, it reminds us of the significance of food for ongoing realizations of autonomy and freedom—not only in African diasporic history—but also in unexpected places such as contemporary urban food deserts, where food-insecure people repurpose public spaces for growing healthy food.

Dependence upon fossil fuels, GMOs, agrochemicals, and machinery sets apart the contemporary era of plantation monocultures in tropical America from its earlier, slave-based incarnation. Nevertheless, oppressive racialized hierarchies continue to bookend the Plantationocene era, even if they appear today in a different guise. This is because the historical and socio-ecological residues of plantation capitalism remain present in current patterns of land use, resource conflict, species extinctions, Black dispossession, and racial discord.

Biocultural refugia stand in sharp contrast to input-intensive monocultures, which contribute to climate change and are vulnerable to economic shocks in world oil and fossil-fuel markets. These multispecies habitats function as living laboratories for innovations in landscapes of tropical food production. Attention to such small-farm food systems, and to the context in which they emerged, enables us to critically think through Plantationocene histories and to craft and enact strategies for food futures that can be equitable, agroecologically sound, and responsive to racial injustices.

Notes

1 Judith A. Carney and Richard N. Rosomoff, *In the Shadow of Slavery: Africa's Botanical Legacy in the Atlantic World* (Berkeley: University of California Press, 2009); David Eltis and David Richardson, *Atlas of the Transatlantic Slave Trade* (New Haven: Yale University Press, 2010); Colleen E. Kriger, *Making Money: Life, Death, and Early Modern Trade on Africa's Guinea Coast* (Athens: Ohio University Press, 2017), 123–25; Angus Dalrymple-Smith and Ewout Frankema, "Slave Ship Provisioning in the Long Eighteenth Century: A Boost to West African Commercial Agriculture?," *European Review of Economic History* 21, no. 2 (2017): 185–235; and Judith A. Carney, "Subsistence in the Plantationocene: Dooryard Gardens, Agrobiodiversity, and the Subaltern Economies of Slavery," *The Journal of Peasant Studies* 48, no. 5 (2021): 1075–99.

2 Ivette Perfecto, John Vandermeer, and Angus Wright, *Nature's Matrix: Linking Agriculture, Conservation and Food Sovereignty* (Sterling, Va.: Earthscan, 2009); and Karl S. Zimmerer and Stef de Hahn, eds., *Agrobiodiversity: Integrating Knowledge for a Sustainable Future* (Cambridge, Mass.: MIT Press, 2019).

3 Charles Kingsley, *At Last: A Christmas in the West Indies* (London: Macmillan, 1871), 115.

4 See, for instance, Johann Jakob Bossart, Arnold R. Highfield, and Vladimir Barac, *C. G. A. Oldendorp's History of the Mission of the Evangelical Brethren on the Caribbean Islands of St. Thomas, St. Croix, and St. John* (Ann Arbor, Mich.: Karoma, 1987); Carney and Rosomoff, *In the Shadow of Slavery*; Case Watkins, *Palm Oil Diaspora: Afro-Brazilian Landscapes and Economies on Bahia's Dendê Coast* (Cambridge: Cambridge University Press, 2021); and Carney, "Subsistence in the Plantationocene."

5 Alfred W. Crosby, *The Columbian Exchange: Biological and Cultural Consequences of 1492* (Westport, Conn.: Greenwood Press, 1972); and Alfred W. Crosby, *Ecological Imperialism: The Biological Expansion of Europe, 900–1900* (Cambridge: Cambridge University Press, 1986).

6 It is estimated that at least one million African captives did not survive the Atlantic voyage across the Middle Passage. Eltis and Richardson, *Atlas of the Transatlantic Slave Trade*.

7 Hans Sloane was the founder of the British Museum who resided in Jamaica at the end of the seventeenth century. Willem Piso and Georg Marcgrave were naturalists on the first (Dutch) scientific expedition to Brazil in the 1640s.

8 William Ed Grimé, *Ethno-Botany of the Black Americans* (Algonac, Mich.: Reference Publications, 1979); and Carney and Rosomoff, *In the Shadow of Slavery*.

9 See, for instance, Bruce L. Mouser, *A Slaving Voyage to Africa and Jamaica: The Log of the Sandown, 1793–1794* (Bloomington: Indiana University Press, 2002); Kriger, *Making Money*; and Pieter de Marees, *Description and Historical Account of the Gold Kingdom of Guinea (1602)*, ed. Albert van Dantzig and Adam Jones (Oxford: Oxford University Press, 1987).

10 Judith A. Carney, *Black Rice: The African Origins of Rice Cultivation in the Americas* (Cambridge, Mass.: Harvard University Press, 2001); Kriger, *Making Money*; and Dalrymple-Smith and Frankema, "Slave Ship Provisioning in the Long Eighteenth Century."

11 Leif Svalesen, *The Slave Ship Fredensborg* (Bloomington: Indiana University Press, 2000), 112; and George Francis Dow, *Slave Ships and Slaving* (Salem, Mass.: Marine Research Society, 1927), 66.

12 Captain Thomas Weaver of the *Swan* to the Royal African Company, May 5, 1704, quoted in David Eltis, *The Rise of African Slavery in the Americas* (Cambridge: Cambridge University Press, 2000), 106n58.

13 Africa's indigenous rice region extends along the coast from present-day Senegal to Ghana and inland to Lake Chad. On women's role on Senegal River slave ships, see James F. Searing, *West African Slavery and Atlantic Commerce: The Senegal River Valley, 1700–1860* (New York: Cambridge University Press, 1993); on Middle Passage crossings, see *Journal of the Slave Ship Mary*, Georgetown University, http://hdl.handle.net/10822/1055276.

14 Smeathman to Drury, Sierra Leone, July 10, 1773, MS D.26, Uppsala University, Sweden. Reference courtesy of Starr Douglas, July 2, 2002.

15 Pinckard, quoted in Dow, *Slave Ships and Slaving*, xxiii–xxiv.

16 M. Vaillant, "Milieu cultural et classification des variétés de riz des Guyanes français et hollandaise," *Journal d'agriculture traditionnelle et de botanique appliquée* 33 (1948): 520–29; Richard Price, *Alabi's World* (Baltimore: Johns Hopkins University Press, 1990); Carney, *Black Rice*; Judith A. Carney, "'With Grains in Her Hair': Rice in Colonial Brazil," *Slavery and Abolition* 25, no. 1 (2004): 1–27; and Tinde van Andel, Harro Maat, and Nicholaas Pinas, "Maroon Women in Suriname and French Guiana: Rice, Slavery, Memory," *Slavery and Abolition* 45, no. 2 (2023): 187–211.

17 Carney, *Black Rice*, 49–55.

18 Richard Price, "Subsistence on the Plantation Periphery: Crops, Cooking, and Labour Among Eighteenth-Century Suriname Maroons," *Slavery and Abolition* 12, no. 1 (1991): 107–27.

19 Tinde van Andel, "African Rice (*Oryza glaberrima* Steud.): Lost Crop of the Enslaved Africans Discovered in Suriname," *Economic Botany* 64, no. 1 (2010): 1–10; Tinde R. van Andel, Rachel S. Meyer, Saulo A. Aflitos, et al., "Tracing Ancestor Rice of Suriname Maroons Back to Its African Origin," *Nature Plants* 2 (2016); and Margaretha A. Veltman, Jonathan M. Flowers, Tinde R. van Andel, and M. Eric Schranz, "Origins and Geographic Diversification of African Rice (*Oryza glaberrima*)," *PlOS One* 14, no. 3 (2019): e0203508, https://doi.org/10.1371/journal.pone.0203508.

20 John Michael Vlach, *The Planter's Prospect: Privilege and Slavery in Plantation Paintings* (Chapel Hill: University of North Carolina Press, 2002); and Elizabeth A. Bohls, *Slavery and the Politics of Place: Representing the Colonial Caribbean, 1770–1833* (New York: Cambridge University Press, 2014).

21 The plantain, an Asian domesticate, formed part of African foodways thousands of years prior to the onset of the transatlantic slave trade. Edmond de Langhe, Ronnie Swennen, and D. Vuylsteke, "Plantain in the Early Bantu World," *Azania: Archaeological Research in Africa* 29, no. 1 (1994): 147–60; and Bryan Edwards, *The History, Civil and Commercial, of the British Colonies in the West Indies*, 2 vols. (London: John Stockdale, 1793), 2:126.

22 George Pinckard, *Notes on the West Indies*, 3 vols. (London: Longman, Hurst, Rees, and Orme, 1806), 1:368.

23 Katherine E. Steir, "Provisions of Power: Food and Scarcity in Jamaica, 1730–1790" (PhD dissertation, Georgetown University, 2021); and Woodville K. Marshall, "Provision Ground and Plantation Labour in Four Windward Islands: Competition for Resources During Slavery," *Slavery and Abolition* 12, no. 1 (1991): 48–67.

24 Dale Tomich, "Une petite Guinée: Provision Ground and Plantation in Martinique, 1830–1848," in *Cultivation and Culture: Labor and the Shaping of Slave Life in the Americas*, ed. Ira Berlin and Philip D. Morgan (Charlottesville: University of Virginia Press, 1993), 222; and Bossart, Highfield, and Barac, *C. G. A. Oldendorp's History of the Mission of the Evangelical Brethren on the Caribbean Islands of St. Thomas, St. Croix, and St. John.*

25 John T. Schneider, *Dictionary of African Borrowings in Brazilian Portuguese* (Hamburg: H. Buske, 1991); Frederic G. Cassidy and Robert Brock Le Page, *Dictionary of Jamaican English* (Kingston: University of the West Indies Press, 2002); and Carney and Rosomoff, *In the Shadow of Slavery*, 139–42.

26 Carney and Rosomoff, *In the Shadow of Slavery*; and Steir, "Provisions of Power," 9.

27 Sidney W. Mintz and Douglas Hall, "The Origins of the Jamaican Internal Marketing System," in *Papers in Caribbean Anthropology*, no. 57 (New Haven: Yale University, Department of Anthropology, 1960).

28 John Gabriel Stedman, *Narrative of a Five Years' Expedition: Against the Revolted Negros of Surinam, in Guiana, on the Wild Coast of South America*, 2 vols. (London: J. Johnson, 1813), 2:128–29; Sylvia W. de Groot, "The Boni Maroon War 1765–1793, Surinam and French Guyana," *Boletín de Estudios Latinoamericanos y del Caribe* 18 (1975): 30–48; and Judith A. Carney and Richard N. Rosomoff, "Covert Cultivars and Clandestine Communities: Rice and the Making of an Afrodescendant Peasantry in Maranhão, Brazil," *The Journal of Peasant Studies* 51, no. 7 (2024): 1626–48.

29 Price, "Subsistence on the Plantation Periphery."

30 Matthias Röhrig Assunção, "Popular Culture and Regional Society in Nineteenth-Century Maranhão, Brazil," *Bulletin of Latin American Research* 14, no. 3 (1995): 265–86; and Flávio dos Santos Gomes, "A Hidra e os pântanos: Quilombos e mocambos no Brasil (séculos XVII–XIX)" (PhD dissertation, University of Campinas, 1997).

31 Kingsley, *At Last*, 309.

32 Darrell A. Posey, *Indigenous Knowledge and Ethics: A Darrell Posey Reader* (New York: Routledge, 2004); and Miguel A. Altieri and Victor Manuel Toledo, "The Agroecological Revolution in Latin America: Rescuing Nature, Ensuring Food Sovereignty and Empowering Peasants," *The Journal of Peasant Studies* 38, no. 3 (2011): 587–612.

33 See, for instance, Riva Berleant-Schiller and Lydia M. Pulsipher, "Subsistence Cultivation in the Caribbean," *New West Indian Guide* 60, nos. 1–2 (1986): 1–40; John S. Brierley, "West Indian Kitchen Gardens: A Historical Perspective with Current Insights from Grenada," *Food and Nutrition Bulletin* 7, no. 3 (1985): 1–10; John S. Brierley, "Kitchen Gardens in the Caribbean, Past and Present: Their Role in Small Farm Development," *Caribbean Geography* 3, no. 1 (1991): 15–28; Theo L. Hills, "The Caribbean Peasant Food Forest: Ecological Artistry or Random Chaos," in *Small Farming and Peasant Resources in the Caribbean*, ed. John S. Brierley and Hymie Rubinstein (Winnipeg: Department of Geography, University of Manitoba, 1988), 1–28; Theo L. Hills and Stanley Iton, "A Reassessment of the 'Traditional' in Caribbean Small-Scale Agriculture," *Caribbean Geography* 1, no. 1 (1983): 24–35; Donald Q. Innis, "The Efficiency of Jamaican Peasant Land Use," *Canadian Geographies* 5, no. 2 (1961): 19–23; and Lydia M. Pulsipher, "The Landscapes and Ideational Roles of Caribbean Slave Gardens," in *The Archaeology of Garden and Field*, ed. Naomi F. Miller and Kathryn L. Gleason (Philadelphia: University of Pennsylvania Press, 1994), 202–22.

34 Michelle Cocks, "Biocultural Diversity: Moving Beyond the Realm of 'Indigenous' and 'Local' People," *Human Ecology* 34, no. 2 (2006): 185–200; Juliana Merçon, Susanne Vetter, Maria Tengö, et al., "From Local Landscapes to International Policy: Contributions of the Biocultural Paradigm to Global Sustainability," *Global Sustainability* 2 (2019), https://doi.org/10.1017/sus.2019.4; and Perfecto, Vandermeer, and Wright, *Nature's Matrix.*

35 Angela May Steward and Deborah de Magalhães Lima, "'We Also Preserve': *Quilombola* Defense of Traditional Plant Management Practices Against Preservationist Bias in Mumbuca, Minas Gerais, Brazil," *Journal of Ethnobiology* 37, no. 1 (2017): 141–65; Maiara Cristina Gonçalves, Fernanda Ribeiro da Silva, Daniele Cantelli, et al., "Traditional Agriculture and Food Sovereignty: *Quilombola* Knowledge and Management of Food Crops," *Journal of Ethnobiology* 42, no. 2 (2022): 241–60; Ludivine Eloy, Cláudia de Souza, Diana Nascimento, et al., "Os sistemas agrícolas tradicionais nos interstícios da soja no Brasil: Processos e limites da conservação da agrobiodiversidade," *Confins: Revue franco-brésilienne de géographie / Revista franco-brasilera de geografia* 45 (2020), https://doi.org/10.4000/confins.28182.

36 Stephan Barthel, Carole Crumley, and Uno Svedin, "Bio-Cultural Refugia—Safeguarding Diversity of Practices for Food Security and Biodiversity," *Global Environmental Change* 23, no. 5 (2013): 1143.

37 Carney and Rosomoff, "Covert Cultivars and Clandestine Communities"; and IBGE (Instituto Brasileiro de Geografia e Estatística), *Censo Demográfico 2022, Quilombolas: Primeiros resultados do universo* (Rio de Janeiro: IBGE, 2023).

3

PALM READINGS

Plantations, Colonialism, and Racial Capitalism in Indonesia and Mexico

PERDANA P. ROSWALDY AND JAYSON M. PORTER

Few oilseeds have influenced the recalibration of racial capitalism, labor migration networks, and uneven economic development during the Age of Imperialism (1870–1914) as much as coconuts and oil palm. Western expansion greased the wheels that turned the palm boom of the Second Industrial Revolution, and palms quickly returned that favor. Energy use tends to define the division between the first, coal-driven Industrial Revolution and the second one fueled by petroleum. However, another source of energy also fed this historical shift: plant fats and oils. While plant fats and oils have held long-standing significance in human history, nineteenth-century developments in chemistry established an epistemic foundation for a global power economy fueled by oleaginous plants. From soap to shortening, processed goods made from coconuts did more than revolutionize industry; they revolutionized the lives of industrial laborers. In the late nineteenth century, palm products increasingly sustained growing urban populations in Amsterdam, New York, and Singapore. Moreover, like sugar in the nineteenth century, coconut and oil palms were colonial adjuncts of the industrial revolutions that reinforced geopolitical disparities between colony and empire.[1] As "biological allies," coconuts also fed colonialism and empire between the United States, Mexico, the Netherlands, and Indonesia.[2]

What was so attractive about coconuts and oil palms, and why were they essential to Western expansion and the Second Industrial Revolution? Although fat is nature's most efficient way of storing energy, there are more books on the power of sugar and wheat than on fats.[3] And while environmental historians of fatty vegetables have demonstrated the myriad ways oilseeds help us tell stories about energy and power, they have yet to fully highlight and connect Latin America and Southeast Asia and their contributions to this botanical history of global power. Colonial logic and hierarchies of race cheapened land and labor in the tropics and justified European extraction. A key argument for capital accumulation and extractivism was that tropical land and labor were only valuable with external capital and ingenuity from temperate zones. As European and American businessmen noticed the local socioeconomic centrality of palms in their tropical colonies, imperialists imagined how capitalism and racism could work together to transform "native" and "backward" ecologies of small-scale, semi-wild, and polycultural plots of coconuts into large-scale, controlled, and monocultural plantations.

Palms illuminate formations and contradictions within colonial and capitalist expansion. Colonial archives, in general, and plantation archives, in particular, are vast, murky, and violent institutions that often require literacy tools or object lessons to best analyze histories and legacies of socioenvironmental change and extraction in the countryside. Reading palms into the lives of people does not adopt a multispecies approach that assumes "a stable human subject to move beyond"; it does not take for granted that many studies of plantations hold "liberal presumptions of the human."[4] By contrast, reading racial and class violence through plants invites a more capacious and dynamic interpretation of humans and the human condition in relation to plantations. Afro-Mexican coconut farmers in one part of Guerrero cannot be reduced to Afro-Mexican coconut farmers in another part of the state, much less to coconut and oil palm farmers in Indonesia. As Vanessa Agard-Jones reminds us, the local matters in analyzing the material cultures and environments of colonial projects.[5]

In their similarities and differences, the histories of palms in Indonesia and Mexico reveal the emergence and dynamics of racial capitalism and modernity. Here, "racial capitalism" refers to how capitalism foregrounds its exploitation in the construction of racial categories. Racial capitalism can dissect plantations and commodities as it illuminates how race and gender are co-constitutive in commodity production and plantations.[6] Analyses of plantations and plants through the lens of racial capitalism help connect the two across history and different places. Plantations perfectly embody how racialized oppression and capital accumulation (re)produce each other.[7] Drawing from archival analysis and ethnography, this study of coconut in Indonesia and Mexico offers a glimpse of the patterns of commodity production generated by plantations across different colonial regimes and historical contexts. We show similarities and differences in operating race, land dispossession, and subjugation to support the empire's project and white supremacy's demand for goods. For example, whether coconut is endemic or alien to the region, it would shape the cross-class and racial solidarity against colonial agribusiness capital. Hence, in addition to its development means and operatives (plantations, racial capitalism), crops also serve as agents of labor and economic change.

Exploring the fortunes of palms in Mexico and Indonesia helps expand the "commodity regime" framework beyond its "regional closets" through the power of plantations as colonial formations that correlate with the persistence of cheap-commodity economies.[8] Palms provide pathways for resource extraction to take off, even as colonialism deployed different approaches in their plantation operatives. We showcase how different commodity histories in Mexico and Indonesia both center on racial capitalism. The racial capitalism here operates as an iteration—including its adaptations—that outlasts different modes of economic systems and labor arrangements across history. In both cases, race is a central component of land and labor relations in plantations. We break from the monocultural framework of a single commodity problem and show how the entanglement of different crops reveals the persistence of racial capitalism inherent in our modern agribusiness and agro-industry.

Indonesia's History of Coconut

The coconut palm (*Cocos nucifera*) grows across the tropical belt. Before European colonialism and racial capitalism spread coconuts worldwide, fossil records indicate that the coconut palm existed in two global centers of palm diversity: Central America and Southeast Asia. The first species of palms began to evolve eighty million years ago, and by fifty-six million years ago, palms were abundant in speciation and widespread in geography, making them some of the best paleo-markers of paleoclimates and climate changes. Palms render paleobiogeography more visible since "the historical separation of landmasses has visible impacts on the distribution of palm species and clades at the interchange areas of Central America and Southeast Asia."[9] Coconut palms can live from sixty to eighty years, and since they do not grow tree rings, we can determine their age by the length of time needed to produce a leaf and by counting leaf scars.

The sun-loving coconut palm flourishes best in coastal ecotones with ample rainfall and sandy soils with good drainage. Easily confused with numerous other palm varieties with "slender, leaning, ringed trunks" and "graceful crowns of giant featherlike leaves," coconut palms stand out as the most economically important of all species of palm in the Arecaceae order.[10] With a seed that contains an average of 1,400 calories and several uses beyond alimentation, from building materials to medicine, *Cocos nucifera* has a long history of attracting human attention and inspiring human imagination. Most paleobotanists agree that humans in modern-day Indonesia or Malaysia first domesticated the coconut palm over 7,500 years ago.[11] Paleo-linguists draw on Sanskrit evidence to estimate that the coconut palm's social status in Asia extends back three to four thousand years.[12] However, the question of how coconuts spread across the tropics and the plant's relationship with human settlement has been debated since the early colonial period. Coconuts crystallized for Europeans how little they knew about the tropical world and, therefore, became a focus of curiosity and extraction around which colonialism and racial capitalism pivoted and expanded.

As seventeenth- and eighteenth-century Spanish and Portuguese slave traders relied on coconuts for drinking water on the ocean, especially for enslaved Africans on board, the colonists began wondering how to best distinguish coconut palms from other, less "useful" palms.[13] Struggling to answer this question, colonialists drew upon preexisting racist assumptions about the capabilities and desires of "primitive" people: they would rely on the biological characteristics of the plant to design their desired labor arrangement to extract the cash crop. From their first encounters with coconuts, Europeans betrayed their extractive interests and engaged in bioprospecting for land and labor. Devaluing the coconut palm as a "lazy man's crop"—because of its low labor input[14]—disparaged Indigenous coconut caretakers without discrediting coconuts as a valuable crop. Colonialists simultaneously saw coconuts as useful and their farmers as useless. With the Second Industrial Revolution and the Age of Imperialism—along with the rise of economic botany and oil chemistry—coconut-related racism gave way to coconut imperialism in the 1880s.[15] Few products embodied this extractive accumulation as well as Lever's Sunlight soap, a product of tropical energy's solidification. Coconuts, thus, served as a colonial and capitalist conduit of energy transformation from solar energy to Sunlight.[16]

The origin of coconuts in Indonesia remains unclear, as the fruit is native to the equatorial region where Indonesia is located. The oldest reference to the coconut is found in the Talang Tuo inscription, originating from the seventh century during the reign of the Sriwijaya Kingdom in present-day South Sumatra. The inscription detailed the fruits (i.e., palm sugar, coconut, betel nuts) that local inhabitants grew in the Seguntang Hill.[17] The Dutch government initially focused on establishing its exploitative grip in the inner islands of Java and Sumatra, where the coconut was not a major commodity, resulting in the late development of coconut cultivation in the nineteenth century. Unlike other colonial commodities at that time, such as tobacco, rubber, and sugar, the coconut is relatively understudied, even as the region was one of the largest exporters of copra (dried coconut kernels for oil production) in the mid-nineteenth century. Coconut production was more dispersed than that of other crops. Furthermore, Dutch colonialism was not as strong in the areas of coconut and copra production, obscuring the crop's importance for the empire.

Extracting copra to satisfy the increasing demand of the oil fats industry in Western Europe made the coconut one of the primary commodities from the East Indies.[18] The unpopularity of the coconut as an analytical device to dissect Dutch colonialism is best explained by the geographical fix of the commodity. Although, in general, coconuts are available throughout the East Indies, cultivation was mostly prioritized in the eastern part of Indonesia (particularly on Celebes/Sulawesi Island), categorized by the Dutch Empire as the *Inlandsche Gemeente Ordonnantie Biutengewsten* or the Municipality Ordinance of the Outer Region.[19] This geographical classification was vital for the colonial government for two reasons: it signified areas that the empire could not yet fully subjugate; and it showcased different approaches used by the colonial government to take over these areas.[20]

Prior to the copra industry, the Indigenous people across Sulawesi had cultivated coconuts primarily for their oil and to barter for cotton and rice. By the late eighteenth century, the island was already exporting copra for coconut oil to Europe as the base for margarine. The inner island of Java had also received substantial coconuts from South Sulawesi since the early nineteenth century, supporting the economy in the eastern East Indies.[21] The Dutch government, however, initially hesitated to integrate the commodity into its more lucrative monoculture/plantation system due to the coconut's long harvest season. The model for coconut cultivation in South Sulawesi relied on garden plots used and worked by Indigenous farmers. The farmers worked on claimless land, sold their harvest, and paid a percentage of their income from the harvest as a tax to the local landed elites.[22] In 1860, when the colonial government finally prioritized coconut trade in the area, the local Dutch office introduced a poll tax that triggered land concentration in the hands of local landed elites. Smaller and common farmers who could not pay the colonial tax had to sell their lots to the local royal and elite families, resulting in massive landlessness among the original coconut cultivators (Figures 3.1 and 3.2).[23]

The Colonial Agrarian Law (*Staatsblad* no. 55) of 1870 stipulated that Natives of the East Indies could not own land, locking them into an exploitative labor relationship orchestrated by the Dutch government but supervised by the local elites. Coconut cultivation and later plantation continuously relied on enslaved people from Maluku and Papua New Guinea further to the east.[24] The Outer Islands—especially where coconut plantations were—received a different amount of political and economic attention from the central colonial government in Batavia, as shown by a few colonial local headquarters stationed in the area.[25] However, the situation only highlighted the importance of the coconut trade, since the eastern headquarters of the Dutch colonial government was located in Makassar, South Sulawesi, the largest producer of coconuts in the colony.

While similar to Mexico in terms of how coconut cultivation became a hotbed for shifting political and economic control, the late colonial state of Indonesia faced a looming crisis in coconut production. In the final years of Dutch colonialism in Indonesia, coconut cultivation and the copra trade set the political stage for power dynamics and grabbing in eastern Indonesia, paving the way for further commodity extraction and expansion post–World War II. Coconut prosperity in the East Indies was relatively short-lived, particularly after the Great Depression and World War II, when cheaper alternatives for edible oils, like *Elaeis* (*guineensis* and *oleifera*) and soybean, were developed. The situation favored oil palm production since the latter's arrival in 1847. In 1940, the colonial government established the Copra Fund, a financial institution whose mission was to rescue and re-boost copra and coconut cultivation in the East Indies.[26] A few years later, however, the Japanese occupation and its Imperial Army halted copra production for cottonseed, almost collapsing the entire copra production. When the Dutch attempted to regain control in the eastern region after Indonesia's independence, copra production was already too diminished to be salvageable.

The Copra Fund's last director general, Henk Peeters, lamented the incompetence of officials from Indonesia's new government and the Dutch Ministry of Agriculture, who were supposed to support the rescue mission.[27] In the last few years before the Dutch occupation ended, the fund managed to triple the copra export. Unfortunately, it only created local inflation during the postwar crisis, and no processing operation could survive.[28] The mismanagement of Indonesia's newly independent government, however, proved the enduring legacy of coconut power in the eastern region: the secessionist rebellion that ensued in 1955 was

FIGURE 3.1

Colonial map of Celebes, 1921. Koninklijk Instituut voor de Tropen, Amsterdam.

FIGURE 3.2

Workers at the Cultuur- en Handel Maatschappij, a Dutch coconut plantation company in Lariang, West Celebes, 1924. Cultuur- en Handel Maatschappij, Lariang.

driven, among other reasons, by the desire to control the coconut economy.

The Story of Coconut in Mexico

A different capital penetration and racialization of croppers are shown in the case of Mexico, where coconuts are not a native species, thus influencing the political economic control and social formation of Mexican coconut production. In Mexico, U.S. and Spanish businesspeople began arriving in Acapulco in the late nineteenth century to establish coconut plantations and factories for the manufacture of soap and vegetable oil. Scholars and scientists have historically debated when, how, and what introduced coconuts to Mexico.[29] Some have argued that the Spanish brought coconuts as a water substitute on slave-trading ships. Others insist that coconuts had precolonial roots; many think that they reached Mexico's Pacific coast with Filipino traders on the annual Manila Galleon to Acapulco.[30] However, the coconut plantation was primarily a foreign transplant that grew from earlier plantation economies—cacao and cotton—before taking over in the early twentieth century.

Before soap factories appeared on the coast in the late nineteenth century, Afro-descendants cultivated and collected a variety of oilseeds for landholders and their own local markets. During the late colonial period, regional *hacendados* (plantation landowners) enslaved Afro-descendants to work in the cotton fields of Costa Chica and Costa Grande.[31] However, after the abolition of slavery, Afro-descendants—now sharecroppers—sustained the coastal cotton economy. Black and Indigenous people also grew coconuts in subsistence plots and *huertas* (small orchards) that reflected their freedom. Coconut and other palms were likely common in *cimarrones* (runaway communities) throughout the colonial period.[32] The coconut palm appears as a complementary crop alongside sesame and fruit trees in these Afro-descendant communities long before the foreign and then national expansion of coconut plantations revived and reinforced colonial hierarchies of control.

When geographer Alfonso Luis Velasco traveled the coast in the 1890s to write his book *Geografía y estadística del estado de Guerrero* (1896), he witnessed "good black [cotton] farmers" in every district of coastal Guerrero from La Unión to Abasolo.[33] Contrary to most scholarship on the region, Luis Velasco observed more Black cotton farmers in Costa Grande than Costa Chica. Luis Velasco did not take photos, but we can imagine a soap factory in Costa Chica run by Black workers. There would be next to no machinery except for a fair deseeder. Black men did the hard labor of crushing the cottonseed, keeping the fire to extract molten oils to mold the soap by hand. Imagine hands hardened by intense heat and heavy lifting: scarred forearms, missing fingers, third- and second-degree burns. There might have been a warehouse building to hold raw cotton that towered over the factory itself. You would have to be on the street to notice the factory because it would not have been a dominant structure one could see from far away. Nevertheless, its impact on the region's economy and ecology would have been felt and visible from much farther away, especially if you were Black, Indigenous, and poor.

As modern and hygienic as it sounds, soap-making relied on extractive plantation economies. In 1896, the Mexican government gave Esteban Benecke of Stephens and Company a concession to establish a vegetable oil and soap factory in Acapulco.[34] They named the factory La Favorita, and by 1899 it produced over one hundred thousand pounds of soap per month from cottonseed, sesame, and cayuco seeds. Black agricultural labor made this possible.[35] Afro-Indigenous women collected seeds; Black, Indigenous, and poor mestizos collected seeds and worked the fields. The soap company began to shift to coconut-based products after Spanish business families purchased the factory the following decade and renamed it La Especial (Figure 3.3).

As the global scramble for coconuts accelerated in the 1880s, politicians in Mexico's Porfiriato (1876–1911) encouraged foreign companies to buy Mexican land cheaply, promising even cheaper labor. The Porfirian period, named after the Mexican dictator Porfirio Díaz, was characterized by widespread government-sanctioned foreign land grabbing predicated on the devaluation of Mexican land and labor, through the

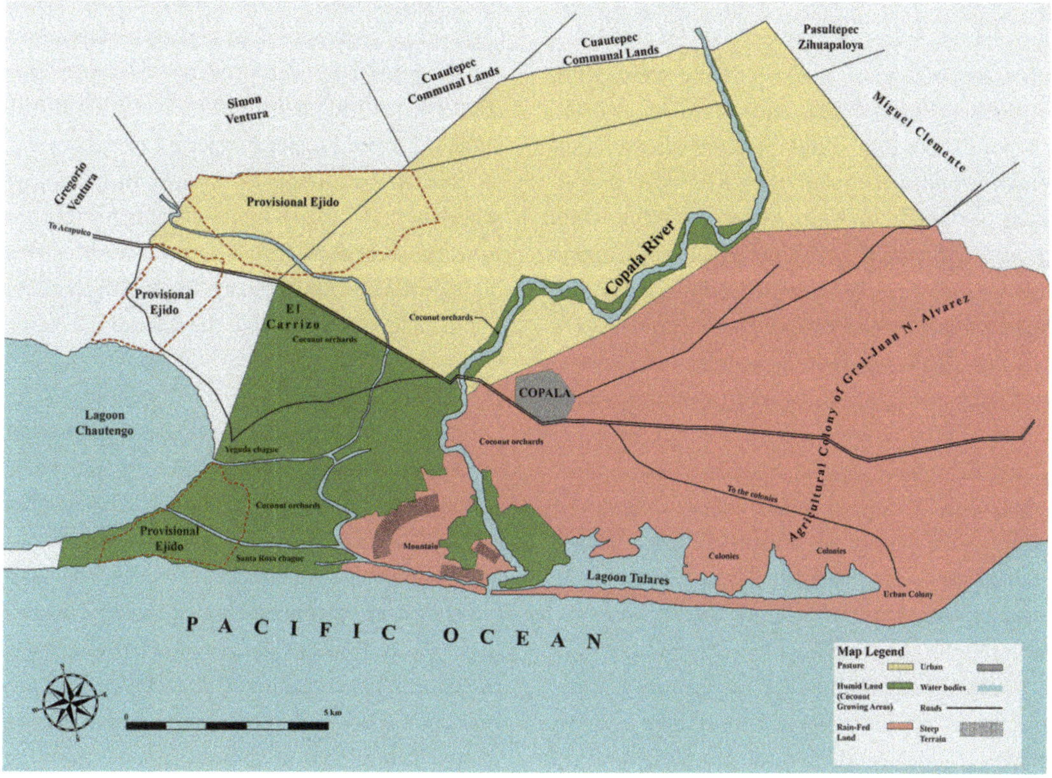

FIGURE 3.3

Coconut-growing regions emerged from cacao- and cotton-growing regions. Map by Alissa Ujie Diamond.

cheapening and sale of mostly Afro-descendant and Indigenous land. The state-sanctioned narrative that Guerrero was economically backward and socially isolated reduced the cost of land while also depicting local labor as abundant, hardworking, and racially diverse. Foreign companies did not hesitate to target the historical Afro-Indigenous coast as a site for coconut expansion.[36] A new palm pathway was built on a legacy of cacao and cotton plantation economies, which relied on Indigenous and Black labor, and began to transform a biodiverse ecosystem of many varieties of palms into a monocultural landscape of the coconut palm. Within decades, coconut expansion almost wiped out other species populations like herons, peccaries, and alligators.[37] This violence came in the form of the Porfirian coconut plantation.

The coconut plantations dominating the landscape were, therefore, products of racial capitalism and Porfirian politics. In the colonial period, mining enterprises and plantations expanded together. The U.S.-owned Guerrero Land and Trading Company purchased 60,000 ha of coastal land in Guerrero in 1907 and quickly started planting coconuts by the thousands. Within four years, the company had replaced sparse and scattered cayuco groves with one hundred thousand coconut palms and three thousand banana trees stretching from Costa Grande to Costa Chica.[38] Around the same time, the Seattle-based Mexican-Pacific Company also purchased 100,000 ha to plant coconut and sesame plantations alongside lime and banana orchards.[39] A hurricane in 1912 devastated a large number of the maturing coconut palms. However, the 800 percent increase in prices for coconuts during World War I helped foreign and Mexican plantation owners bounce back by the end of the Mexican Revolution (1910–1920).[40] This early craze for coconuts was not restricted to U.S. companies; the 150 or so of the largest Spanish and Mexican landowners in the region also began investing in coconuts. Little did landowners realize that encouraging sharecroppers to plant coconuts after the Mexican Revolution would temporarily aid local struggles for agrarian reform on the coast.

The spread of coconut palms bolstered landed power but also had an oddly democratizing impact on the coast, which became increasingly problematic for landowners after the Mexican Revolution. Since the commercial value of coconuts was not going down anytime soon, landowners encouraged their tenant farmers to plant coconuts. Tenants could plant small orchards on their rented land and pay a fixed rent of one peso per tree.[41] After five years of tending their groves, the tenant could earn two pesos annually for each tree. This practice did not guarantee sharecroppers ownership of the coconut groves but instead allowed tenants to profit from temporary palm ownership. Coconuts meant real money for some poor farmers, especially in the lowlands where Afro-descendants have grown coconuts for generations. An ambitious and skilled farmer could always have their grove usurped by a landowner wanting to start or extend a coconut plantation. At the same time, landowners could evict "improving" tenants who increased the value of their holdings.

After the Mexican Revolution, access to coconut groves became a central concern of agrarian reform in coastal Guerrero. Landless peasant farmers hoped to establish their *ejidos* (collectively held land grants) around palms but could not do so without conflict with landowners and smallholders. During the coconut boom that coincided with World War II, many communal *ejidos* were co-opted by agribusiness, foreign and domestic. Connected to multinational companies like Procter and Gamble and Bola de Nieve, coconut plantations came to dominate the landscape; they did so in uneven, unexpected, and unfortunate ways. By 1960, Guerrero was the largest coconut producer in the Western Hemisphere, but few coconut farmers along the 600 km coast of plantations benefited from the enterprise.[42]

The Afterlife of Coconut Cultivation and the Next Green Gold

While coconut cultivation flourished in the nineteenth century, global cash crop management and the colonial political economy in the early twentieth century altered its production. Waves of liberal policies forced colonial governments to adjust their agricultural management under the pretext of modernization and a benevolent empire.[43] When the Dutch colonial government issued the *Agrarische Wet* (Agrarian Law) in 1870, it ended the Forced Cultivation System, only to replace it with monoculture/plantations and prioritize the latter. This shift also marked the so-called liberal era of the Dutch East Indies, which referred to the increasing volume of foreign investment to fund colonial plantation expansion, particularly for tobacco, palm oil, and rubber. Similar liberal land reforms during the Mexican Porfiriato also laid the path for U.S.-owned companies like the Guerrero Trading Company to buy 465,000 ac in 1902 to develop coconut plantations.[44]

Disciplining coconut cultivation to support the empire required changes in both land governance and labor relations, tailored explicitly for plantations. Now that slavery was technically outlawed, colonial governments reinforced racial hierarchies imbued in colonial agriculture by new means.[45] Technological change in oilseed industrialization, from machinery to chemistry and processing, helped to reinforce racial hierarchies by marginalizing labor in the tropics and centering temperate whiteness. Calling such a change "technological advancement" without reckoning with how it advances white supremacy helps normalize how capitalism benefits from racism and vice versa. Consider the rise of drying centers for the driest, whitest industrial copra. Producers felt pressured to dry copra further as demand rose in the international market. The drier the copra, the harder it was to spoil, and the easier it was to extract oil. The whiteness of very dry copra also gave the added illusion of purity. The will for whiteness led to a gradual shift from sun-drying copra to drying centers. Arguments against the reliability of the tropical climate helped to hasten the technological change; drying centers are also more dependable in regions with heavy or sporadic rains. Poor farmers drying their copra when weather permitted could not compete with wealthier farmers or hoarders drying copra in centers.[46] Moreover, hoarders with drying centers could hold onto products even longer and wait for the best prices at the expense of farmers.

Whereas the industrial processing of oilseeds undermined profits for poor farmers, advertising processed goods has a long-standing history of promoting whiteness. Despite distinct chemical and botanical differences between coconuts and other oilseeds, companies use technology to neutralize differences and make different oilseeds fungible parts in finished goods. The interchangeability of copra, cotton, sesame, and other oilseeds aided coconut consumption and fostered a culture of overproduction that adversely affected all oilseed farmers. For coconut farmers in Indonesia and Mexico, nothing seemed neutral about this process; living legacies of colonialism, slavery, and resistance animated this slow and structural violence of the coconut economy.[47] The way companies advertised coconut-based processed goods, especially soap products, was also deeply colonial and racist. Through politicized and racialized concepts, namely hygiene and moral cleansing, politicians aimed to surveil, criticize, and control poor people through sanitary campaigns and codes. Whiteness as cleanliness was the most common trope in selling soap: marketing campaigns for Ivory, Snowball, Procter & Gamble, and Bola de Nieve sold whiteness.[48] Advertisements like the one for Pears' Soap illustrated the whitening power of soap: saponification is whiteness as witness (Figure 3.4).

FIGURE 3.4 Advertisment for Pears' Soap, published in *Ainslee's Magazine*, October 1899.

The processed sanitary goods from coconuts under this new regime of colonial plantations also set whiteness as a moral compass: the mandatory and racist call for sanitary conditions in the colony set up the whiteness/whitening process as moral cleansing.[49] Cultivating crops in plantations thus became a holy crusade for many (white) European and American planters, a duty that only the chosen (white men) could fulfill, giving them the right to throw racist slurs and assert moral superiority. The new system of commodity production in plantations required different means of discipline, as a failed attempt to obscure the racialization and subjugation of colonial subjects. This does not mean that the hierarchy was not visible; rather, colonial governments found loopholes to continue to enforce enslavement and forced labor despite the waves of liberalization. Instead of directly legitimizing slurs and other eugenic features of colonial subjects, the empire created racial hierarchies based on production mechanisms and labor stereotypes attached to different races and ethnicities in the colonies.[50]

The coconut frenzy in both Indonesia and Mexico set off a new wave of commodity expansion for the next green gold: palm oil. Coconut cultivation proved a difficult task for the colonial power, which found it harder to fully subjugate its subjects due to the nature of this "lazy man's crop" that required less labor and capital than other, more profitable oilseeds, such as palm oil. Furthermore, coconut cultivation had a democratic impact in different ways for Indonesia and Mexico despite the racial and labor hierarchies. Coconuts financially aided the smallholders and sharecroppers of the East Indies and Mexico, creating an anti-colonial alliance between landed elites and new small farmers for pricing protection and agrarian struggles.[51] Simultaneously, the global commodity market shifted and favored palm oil produced in the British African colonies, decreasing the price and eliminating palm oil as a luxury item. Hence, coconut cultivation became politically and economically less appealing for the colonial government to maintain order and power. Coconut's "lazy man's crop" moniker could not fit the new empire's methods of subjugation, especially under the new plantation regime. Transforming and expanding to palm oil justified different labor systems and land extraction, albeit still using the same logic of discipline as in previous cash crop cultivation.

Toward the late twentieth century, coconut cultivation became a less popular agribusiness despite its history of stable production in Mexico and Indonesia. Soy, corn, and palm oil became the most important "green gold" crops in postcolonial Global South development. The massive turn to palm oil started in the late 1970s in Indonesia.[52] Large amounts of international agricultural loans and aid specifically funded plantation-based labor migration and promoted smallholding to eradicate rural poverty. Hence, it was necessary to support readily available cash crops with a more industrialized production mechanism and faster investment return, in which palm oil fits better than coconut.

The transition to palm oil,[53] particularly in Indonesia, is an economic choice and a deliberate social, technological, and political decision. Palm oil demands more scrutinized agricultural operations than coconut palm and provides a more lucrative price. In 1848, Dutch botanists brought the plant to Indonesia, where it initially served as a decorative plant for the Colonial Botanical Garden.[54] In its early cultivation, the colonial government discovered that naturalized oil palm in Indonesia produced more oil than its original West African seed, increasing the production, supply, and demand for oil fat.[55] With the Green Revolution in Indonesia in the 1970s, plantations and palm oil became synonymous with development and modernity, a sentiment that has been shared and reproduced by many.[56] Sakreh, a smallholder in Riau, confessed: "Everybody is sick of being poor and backward in the village, and the only way out of that is through palm oil."[57]

Many smallholders and farmers also perceive palm oil as more lucrative than coconuts, even when coconuts demand less capital input than palm oil. Unlike its oil palm counterpart, Indonesia's copra and coconut production rarely receives the state's political and financial protection. Coconut is also less visible in Indonesia's political sphere: hence, the understudy and underappreciation of the commodity. State officials also tend to act harsher toward coconut farmers and production companies, as the crop is not their top economic priority or because their province is en route to replacing coconuts with palm oil: "Ms. Yanti, District B's Local Agricultural Office's vice coordinator, recently requested to revoke a coconut plantation company's permit. She believed it was necessary because the company had not planted anything for over two years. I pointed out that many palm oil plantation companies have worse underperformance and financial losses. Yet, Ms. Yanti never issued a warning to palm oil companies. 'We prefer talking to the companies first. After all, it's palm oil. We need the palm oil money more [than the coconut].'"[58] Even as concerns over palm oil sustainability are at an all-time high following the new wave of international green policies, funding, and development programs, palm oil reigns supreme among other cash crops in Indonesia and Mexico. In the 2022 Roundtable Sustainable Palm Oil annual conference, for example, a representative of Pepsi expressed their interest in involving more Mexican smallholders for sustainable certification, as Mexico remains their most crucial palm oil supply market in North America.[59] The same goes for Indonesia, which, despite going against the European Union Deforestation Regulation, insists that more companies and smallholders comply with sustainability and no-deforestation principles to access climate funds related to Good Agricultural Practices.

Vast areas of land where Indigenous and local people used to cultivate varieties of palms (coconuts, sago, and sugar) have been replaced by meticulously organized palm oil trees. This time, the expansion uses the language of "sustainability" and "renewable energy." Sukardi, a farmer in Riau, just lost his coconut slots due to land grabbing by a palm oil company. "They do anything for palm oil, this province,"[60] he snickered, "even someone's coconut farm that they don't even bother to know or care about." Sukardi's slots are near the coastal area; his district frequently suffers from haze due to palm oil plantations that unlawfully stand on the top of protected peatland. In 2023, the Indonesian government ruled that palm oil companies already operating in the peatland can access replanting funds, furthering the transition to the second green gold.

Plantations Are Colonialism

From Indonesia to Mexico, across the tropics' coconut and oil palm belt, palm plantations masquerade as forests while they suffocate land, labor, and other species. As colonial units of capital and mechanisms of capital accumulation, plantations actively prioritize profits by reducing the value and visibility of place and people.[61] Racist modes have always informed this expression of colonial ethnocentrism, theft, and violence, from displacement to enslavement and enclosure. No matter how "advanced" plantations become in the twenty-first century—through technology, science, or socioeconomic schemes—they will continue to reflect and advance previous colonial metrics, logistics, and strategies. Such durability may be explained by the rigid land and labor relations on which plantations depend for extraction. No matter how advanced capitalism becomes, it will always rely on the ruins of colonial plantations.[62] Ironically, however, such "advancement" has become another myth of progress promised by plantations, as many workers, smallholders, and scholars find that plantations still use low technology. If anything, the advancement is made in the plantation's mode of exploitation. As Udin, the smallholder, noted: "I think the company just gets more creative with their thievery."[63]

Oilseed plantations served as the mechanism of colonial order and progress transition in Indonesia and Mexico from 1880 to 1932. With different intentions and rulers, plantations have continuously generated violence, deepened racial segregation, and maintained inequality. If plantations are, indeed, a "capitalist invention par excellence," they were not made from nothing. Through plantations, capitalism shares colonialism's material culture and socioeconomic landscape. Across the global palm plantation belt, capitalism grew directly from the toxic colonial soils. Saying plantations are colonialism highlights the continuation and differentiation of world-making that colonialism initiated through plantations.[64] Plantations are not just colonial capitalist relics or by-products of despotic postcolonial regimes unable to escape their colonial specters. Plantations can be postcolonial, and many states navigate plantations as tools for reordering their "postcolonial" societies. Plantations are the colonial past and present; they exist beyond a temporal stamp in our history.[65]

Plantations "strip away the complexities of land" by deliberately suppressing workers and breaking the totality of land and ecosystems from other landscapes (rivers, forests, and Indigenous communities).[66] This is a call to understand how multiple nodes of oppression alter plantations and commodity production, contributing to our contemporary world-making plagued by socio-environmental crises. Patriarchy, for instance, feeds plantation expansion by simplifying decision-making to a monoculture of elite men who actively devalue the ideas and contributions of women and poor people of color.[67] Plantations need suitable bodies at the right time in the right place; hence, disciplining labor and land in plantations involves not only land grabbing and economic exploitation but also patriarchy, ableism, and racism.[68] All of these forces are embedded in plantations and perpetuate commodity frontiers. Situating plantations as colonialism simultaneously interrogates plantation violence as colonial while underscoring its adaptability in forming new epochs of social and environmental crises.

Notes

1 Sidney W. Mintz, *Sweetness and Power: The Place of Sugar in Modern History* (New York: Penguin Books, 1986).
2 Alfred W. Crosby, *Ecological Imperialism: The Biological Expansion of Europe, 900–1900* (Cambridge: Cambridge University Press, 1986).
3 Jonathan Robins, "Oil Boom: Agriculture, Chemistry, and the Rise of Global Plant Fat Industries, ca. 1850–1920," *Journal of World History* 29, no. 3 (2018): 313–42.
4 Alyssa Paredes, "Plantation Liberalism," *Current Anthropology* (forthcoming).
5 Vanessa Agard-Jones, "What the Sand Remembers," *GLQ: A Journal of Lesbian and Gay Studies* 18, nos. 2–3 (2012): 326–27.
6 Katherine McKittrick, "Plantation Futures," *Small Axe: A Journal of Criticism* 17, no. 3 (2013): 3–4; Sylvia Wynter, "Novel and History: Plot and Plantation," in *We Must Learn to Sit Down Together and Talk About a Little Culture: Decolonizing Essays, 1967–1984* (Leeds: Peepal Tree Press, 2012); and George L. Beckford, "Caribbean Peasantry in the Confines of the Plantation Mode of Production," *International Social Science Journal* 37, no. 3 (1985): 401–14.
7 Kris Manjapra, "Plantation Dispossessions: The Global Travel of Agricultural Racial Capitalism," in *American Capitalism: New Histories*, ed. Sven Beckert and Christine Desan (New York: Columbia University Press, 2018); Michael Ralph and Maya Singhal, "Racial Capitalism," *Theory and Society* 48, no. 6 (2019): 851–81, https://doi.org/10.1007/s11186-019-09367-z; and Laura Pulido, "Geographies of Race and Ethnicity II: Environmental Racism, Racial Capitalism, and State-Sanctioned Violence," *Progress in Human Geography* 41, no. 4 (2017): 524–33, https://doi.org/10.1177/0309132516646495.
8 McKittrick, "Plantation Futures"; Wynter, "Novel and History, Plot and Plantation"; and Mythri Jegathesan, "Black Feminist Plots Before the Plantationocene and Anthropology's 'Regional Closets,'" *Feminist Anthropology* 2, no. 1 (2021): 78–93.
9 Wolf L. Eiserhardt, Jens-Christian Svenning, W. Daniel Kissling, and Henrik Balslev, "Geographical Ecology of the Palms (Arecaceae): Determinants of Diversity and Distributions Across Spatial Scales," *Annals of Botany* 108, no. 8 (2011): 1405.
10 John Kress, "Coconut Palm (*Cocos nucifera*)," *Encyclopedia Britannica*, June 6, 2025, https://www.britannica.com/plant/coconut-palm.
11 B. K. Maloney, "Palaeoecology and the Origin of the Coconut," *GeoJournal* 31, no. 4 (1993): 355–62.
12 E. H. Man, "A Brief Account of the Nicobar Islanders, with Special Reference to the Inland Tribe of Great Nicobar," *The Journal of the Anthropological Institute of Great Britain and Ireland* 15 (1886): 359.
13 It is worth noting that European slave traders used coconuts and oil palms to feed enslaved Africans during the Middle Passage. Case Watkins, *Palm Oil Diaspora: Afro-Brazilian Landscapes and Economies on Bahia's Dendê Coast* (Cambridge: Cambridge University Press, 2021), 69; and Constance L. Kirker and Mary Newman, *Coconut: A Global History* (London: Reaktion, 2022).
14 Christiaan G. Heersink, "Selayar and the Green Gold: The Development of the Coconut Trade on an Indonesian Island (1820–1950)," *Journal of Southeast Asian Studies* 25, no. 1 (1994): 47–69.
15 Holger Droessler, *Coconut Colonialism: Workers and the Globalization of Samoa* (Cambridge, Mass.: Harvard University Press, 2022).
16 Timothy Burke, "'Sunlight soap has changed my life': Hygiene, Commodification, and the Body in Colonial Zimbabwe," in *Clothing and Difference: Embodied Identities in Colonial and Postcolonial Africa*, ed. Hildi Hendrickson (Durham: Duke University Press, 1996), 189–212; and Anne McClintock, *Imperial Leather: Race, Gender, and Sexuality in the Colonial Contest* (New York: Routledge, 1995).
17 George Cœdès, "Les Inscriptions Malaises de Çrīvijaya," *Bulletin de l'Ecole Française d'Extrême-Orient* 30 (1930): 29–80.
18 Pierre van der Eng, "Food for Growth: Trends in Indonesia's Food Supply, 1880–1995," *Journal of Interdisciplinary History* 30, no. 4 (2000): 591–616; and Pierre van der Eng, "Exploring Exploitation: The Netherlands and Colonial Indonesia 1870–1940," *Revista de historia economica / Journal of Iberian and Latin American Economic History* 16, no. 1 (1998): 291–321.
19 Ulbe Bosma, *The Making of a Periphery: How Island Southeast Asia Became a Mass Exporter of Labor* (New York: Columbia University Press, 2019); Ann Laura Stoler, *Capitalism and Confrontation in Sumatra's Plantation Belt, 1870–1979* (Ann Arbor: University of Michigan Press, 1995); and Yando Zakaria, *Abih Tandeh: Masyarakat Desa Di Bawah Rejim Orde Baru* (Jakarta: ELSAM, 2000).
20 On the other hand, the Municipality Ordinance of the Inner Region (*Inlandsche Gemeente Ordinantie*) only refers to the islands of Java and Madura, where the Dutch first arrived and stationed their central colonial government. Many historians and scholars of Indonesia and Dutch colonialism believe this was the first historical event correlating with Indonesia's contemporary uneven development.
21 Nicholas J. White, "Surviving Sukarno: British Business in Postcolonial Indonesia, 1950–1967," *Modern Asian Studies* 46, no. 5 (2012): 1277–1315.
22 Christiaan Heersink, *Dependence on Green Gold: A Socio-Economic History of the Indonesian Coconut Island Selayar* (Leiden: KITLV Press, 1999).
23 Robert Bridson Cribb, ed., *The Late Colonial State in Indonesia: Political and Economic Foundations of the Netherlands Indies, 1880–1942* (Leiden: KITLV Press, 1994).
24 Jan Pouwer, "The Colonisation, Decolonisation and Recolonisation of West New Guinea," *The Journal of Pacific History* 34, no. 2 (1999): 157–79; and Ligia J. Giay, "Papuans Within Slavery, Slavery Within Papuans: Traces of the Slavery Past in New Guinea, 1600s–1950s" (MA thesis, Universiteit Leiden, 2018).

25 A. H. P. Clemens and J. Thomas Linblad, *Het Belang van de Buitengewesten: Economische Expansie en Koloniale Staatsvorming in de Buitengewesten van Nederlands-Indië, 1870–1942* (Amsterdam: NEHA, 1989).
26 Howard Dick, *The Emergence of a National Economy: An Economic History of Indonesia, 1800–2000* (Honolulu: University of Hawai'i Press, 2002).
27 Willem Peeters, "De tot mislukking gedoemde missie van een fatsoenlijk koloniaal," Historiek, February 1, 2016, https://historiek.net/mislukking-gedoemde-missie-fatsoenlijk-koloniaal/56628/.
28 G. A. Kessler, "Indonesische kroniek," *De Economist* 100, no. 1 (1952): 43–52, https://doi.org/10.1007/BF02287609.
29 Daniel Zizumbo-Villarreal, F. Hernández Roque, and H. C. Harries, "Coconut Varieties in Mexico," *Economic Botany* 47, no. 1 (1993): 65–78; and Charles R. Clement, Daniel Zizumbo-Villarreal, Cecil H. Brown, et al., "Coconuts in the Americas," *The Botanical Review* 79, no. 3 (2013): 342–70.
30 Henry J. Bruman, "Early Coconut Culture in Western Mexico," *Hispanic American Historical Review* 25, no. 2 (1945): 212–23.
31 Gonzalo Aguirre Beltrán, *Cuijla: Esbozo etnográfico de un pueblo negro* (Mexico City: Fondo de Cultura Económica, 1958).
32 Alfonso Luis Velasco, *Geografía y estadística del estado de Guerrero* (Mexico City: Oficina Tipográfica de la Secretaría de Fomento, 1896).
33 Velasco, *Geografía y estadística del estado de Guerrero*.
34 *Boletín de agricultura, minería e industrias* 5, no. 7 (January 1896): 143.
35 Clement S. Edwards, "Commerce in Southern Mexico: Acapulco," *Daily Consular and Trade Reports* (August 16, 1913): 929–30.
36 Francisco A. Gomezjara, *Aceites, jabones y multinacionales* (Mexico City: Ediciones Nueva Sociología, 1978); and Ian Jacobs, *Ranchero Revolt: The Mexican Revolution in Guerrero* (Austin: University of Texas Press, 1982).
37 Epigmenio López Barroso, *Diccionario geográfico, histórico y estadístico del distrito de Abasolo, del estado de Guerrero* (Mexico City: Ediciones Botas, 1967), 74–75. All consular reports noted a few thousand pounds of non-cattle hides from various animals.
38 Consul Marion Letcher, "Acapulco," in *Commercial Relations of the United States with Foreign Countries* (Washington, D.C.: Government Printing Office, 1911), 517.
39 The company imported the bananas from Costa Rica and Panama. Consul Maxwell K. Moorehead, "Acapulco," in *Commercial Relations of the United States with Foreign Countries During the Year 1908* (Washington, D.C.: Government Printing Office, 1909), 141.
40 Otto Wilson, *Oils, Fats, and Waxes in Latin America* (Washington, D.C.: Government Printing Office, 1924), 5.
41 Letcher, "Acapulco," 517.
42 "México, cuarto productor de la copra durante el año de 1958," *El Universal*, May 9, 1959, Biblioteca Miguel Lerdo de Tejada, PO5010, Copra, Producción, ff. 47.
43 Julian Go, "The Provinciality of American Empire: 'Liberal Exceptionalism' and U.S. Colonial Rule, 1898–1912," *Comparative Studies in Society and History* 49, no. 1 (2007): 74–108; and Syed Hussein Alatas, *The Myth of the Lazy Native: A Study of the Image of the Malays, Filipinos and Javanese from the 16th to the 20th Century and Its Function in the Ideology of Colonial Capitalism* (London: F. Cass, 1977).
44 *Historia de la cuestión agraria mexicana: Estado de Guerrero, 1867–1940* (Mexico City: Gobierno del Estado de Guerrero, 1987).
45 Bosma, *The Making of a Periphery*, found that slavery existed in the Dutch East Indies at least until the 1930s. Slavery transformed into coerced labor, indentured/bondage labor, prison labor, coolies, and racialized housing for workers on plantations.
46 P. S. Tilson, "Copra and Its Products," *The Cotton Oil Press* 2, no. 1 (1918): 40–41; and Katharine Snodgrass, *Copra and Coconut Oil* (Stanford, Calif.: Food Research Institute, 1928), 20.
47 Jayson M. Porter, "Making the Coast Pacific: Oilseeds, Environmental Violence and Justice in Guerrero and Sinaloa" (PhD dissertation, Northwestern University, 2022).
48 Helen Zoe Veit, "Eating Cotton: Cottonseed, Crisco, and Consumer Ignorance," *The Journal of the Gilded Age and Progressive Era* 18, no. 4 (2019): 397–421.
49 Marieke Bloembergen, "Being Clean Is Being Strong: Policing Cleanliness and Gay Vices in the Netherlands Indies in the 1930s," in *Cleanliness and Culture: Indonesian Histories*, ed. Kees van Dijk and Jean Gelman Taylor (Leiden: Brill, 2011), 117–45.
50 Jan Breman, *Taming the Coolie Beast: Plantation Society and the Colonial Order in Southeast Asia* (New York: Oxford University Press, 1989); and Vincent J. H. Houben and J. Thomas Lindblad, *Coolie Labour in Colonial Indonesia: A Study of Labour Relations in the Outer Islands, c. 1900–1940* (Wiesbaden: Otto Harrassowitz Verlag, 1999).
51 Heersink, "Selayar and the Green Gold."
52 Since 2008, the Mexican government and Petróleos Mexicanos (PEMEX) have promoted oil palm expansion in Chiapas to diversify sustainable energy in rural Mexico. Yolanda Cristina Massieu Trigo and Blanca Olivia Acuña Rodarte, "Biofuels: Socio-Environmental Violence in Rural Mexico," *Latin American Perspectives* 42, no. 5 (2015): 67–82.
53 One of the largest palm oil companies in Indonesia, Sinar Mas Group, started as a copra-processing business.
54 Luthfi Adam, "Cultivating Power: Buitenzorg Botanic Garden and Empire-Building in the Netherlands East Indies, 1745–1917" (PhD dissertation, Northwestern University, 2020); and Watkins, *Palm Oil Diaspora*, 1, 70.
55 Jonathan E. Robins, *Oil Palm: A Global History* (Chapel Hill: University of North Carolina Press, 2021).
56 Tania Murray Li and Pujo Semedi, *Plantation Life: Corporate Occupation in Indonesia's Oil Palm Zone* (Durham: Duke University Press, 2021); Alice Baudoin, Pierre-Marie Bosc, Cécile Bessou, and Patrice Levang, "A Short History of Palm Oil in Indonesia: A 'Success Story,'" *Review of the Diversity of Palm Oil Production Systems in Indonesia* (2017), https://www.jstor.org/stable/resrep16293.4.

57 Perdana P. Roswaldy, ethnographic field notes, October 19, 2021.
58 Roswaldy, ethnographic field notes, March 20, 2023.
59 Roswaldy, ethnographic field notes, November 30, 2022.
60 Riau is Indonesia's largest palm oil–producing province, with the most significant palm oil plantations. Roswaldy, ethnographic field notes, October 21, 2021.
61 George L. Beckford, *Persistent Poverty: Underdevelopment in Plantation Economies of the Third World*, 2nd ed. (Kingston: University of West Indies Press, 1999); and Katherine McKittrick, "On Plantations, Prisons, and a Black Sense of Place," *Social & Cultural Geography* 12 (2011): 947–63.
62 Major Samuel M. Harrington, "The Strategy and Tactics of Small Wars," *Marine Corps Gazette* 7, no. 1 (1922): 84–93. To invert and paraphrase Samuel Harrington on technology, empire, and surveillance: In small wars, air service will do much to relieve the infantry of its labors. But no matter what means of warfare modern science may invent, the man on foot must, in the last analysis, drive out the man on foot and seize and hold land because mankind lives on the earth. Infantry will therefore always be of importance.
63 Roswaldy interview with Udin, October 14, 2022.
64 Inspired by Liboiron, see Max Liboiron, *Pollution Is Colonialism* (Durham: Duke University Press, 2021).
65 See Black environmentalist and feminist critiques of "Plantationocene" under notes 6 and 8 above, and Jason Moore's explanation to avoid using "Capitalocene" as a temporal remark.
66 Liboiron, *Pollution Is Colonialism*, 80.
67 For more on patriarchy and monoculture, see Amalia Leguizamón, *Seeds of Power: Environmental Injustice and Genetically Modified Soybeans in Argentina* (Durham: Duke University Press, 2020).
68 Tania Murray Li, "Intergenerational Displacement in Indonesia's Oil Palm Plantation Zone," *The Journal of Peasant Studies* 44, no. 6 (2017): 1158–76; Tania Murray Li, *Social Impacts of Oil Palm in Indonesia: A Gendered Perspective from West Kalimantan* (Bogor: Center for International Forestry Research, 2015); and Miranda Morgan, "Women, Gender and Protest: Contesting Oil Palm Plantation Expansion in Indonesia," *The Journal of Peasant Studies* 44, no. 6 (2017): 1177–96.

4

MIGRATIONS

Tales of Uprooting, Transplanting, and Grafting

GIOVANNI ALOI

The Secret Lemon Grove

We've been driving for almost an hour. The sky darkens so early. Days are meant to get longer right after the winter solstice, but February always seems oddly out of sync. It must be the cold. We're heading southwest in a rattly van to pick up a table bought on Facebook Marketplace for Chris's office. He's thrilled. Sleek, modernist, cherry-stained teak, foldable. A steal.

Very few things are as drab and depressing as the land trapped between the 55 and 290—a relentlessly flat and seemingly endless loop of billboards and neglected plazas. We chat and play music while I distractedly scroll Marketplace, hoping for a surprising last-minute find in the area. A few seconds later, a gorgeous driftwood coffee table pops up on the screen. It's a mesmerizing tangle of roots, cut out in the shape of a square. I tell Chris: "Hey, look at this! It's 3 feet × 3 feet × 1.5 feet. Do you think it'd fit by the sofa? $400. It's just a 10-minute drive south of our first stop." We want it. Deal. I message the seller, knowing that the chances of a quick reply are very slim on a late Sunday afternoon, but Hayat is quick: "Yes, we're home! Cash only."

We arrange to swing by between 4:30 and 5:00. We collect the first table, then stop on the way to the second seller to withdraw some cash. The GPS takes us down some country lanes. It's pitch-black. The houses seem nice—middle class? The road gets windier; more homes appear. The soulless, metallic GPS voice announces: "You've arrived! Your destination is on the left." We see the house, but nobody's at the front door. Hayat sends another text: "Pull up to the garage door." We do and get out of the car. It's freezing—one of those days when the high temperature never dares to crawl above 30°F. I still can't see anyone. Then, all of a sudden, the garage door lifts slowly. It rattles and squeaks like any other garage

FIGURE 4.1
Giovanni Aloi, *Lemon Blooms*, 2019.

door in the United States. Warm light spills onto the driveway. The door pulls up completely with a dull thud . . . and we just can't believe our eyes.

Inside, there are no cars, no workbenches cluttered with scattered tools, no stacks of boxes heaped against a single wall. Not even a lawn mower . . .

The garage is filled with trees—seven- to ten-foot-tall lemon trees in massive pots. They are in bloom. The scent pierces through the frozen air; it's heavenly. Dark green, waxy leaves shimmer under the electric light. I can't see the table we've come to collect, but right now it doesn't matter. There's so much unexpected beauty to take in at once. I try to process what I see; the sheer incongruity of this vision clashes with the idea of a suburban Chicago setting in the middle of winter on a cold Sunday afternoon (Figure 4.1).

The garage is attached to the house. Hayat comes through the side door. He's young, early twenties? Handsome. Long, dark hair. He says "hi" and points at the table: "There it is."

"Yes! I see. Thanks. Looks great. Just like the pictures online."

But what I really want to know is how this "garden-garage" came to be. It is not just the lemon trees in bloom. All around, there are hundreds of pots of all sizes lined up on shelves. Mostly plastic, the odd terracotta ones, nothing fancy. Some have plants in them—hibiscuses, geraniums, and elephant ears. Others only seem to contain soil or just the remainders of summer growth, wilted—most likely the bulbs of cannas, overwintering.

I would hate to be rude. It certainly isn't kind to ask about the contents of someone's garage, but I just could not leave without knowing more. "To hell," I think: "I'm buying a table, and I won't ever see him again . . . it will be ok."

"I don't mean to be nosy, but whose plants are these?"

The side door opens again, and a man walks out. He is older than all of us and shorter than me. He waves at us but doesn't say a word.

Hayat says: "Oh, yes. The plants? They belong to my father! That's him. Hey, Dad."

The man waves and smiles. They are both kind and welcoming, but they can't linger because the cold is bad for the plants. So I push my luck and ask one more question, knowing it will be my last: "Do you take all these plants out in the yard in the summer?" "Yes," the elderly man says. Hayat continues: "We are from Iran. My dad moved here in his twenties. These are the plants from his childhood. Every spring, we take them all out in the garden, and they spend winter here." The smile on the elderly man's face widens. Hand on heart: "Oh my! I do the same myself. But my plants go into the basement because our garage is detached, and it gets too cold for them." Hayat translates into Iranian for his dad. The elderly man smiles back. I glance once more at the incredible garage-garden, trying to take in the outstanding view. I can't possibly ask to take a picture. It would be so tactless and utterly crass. I bite my tongue. I look at Chris: "Let's get that table into the van." It's so heavy.

As we load it, a thread begins to unravel—two souls, both immigrants, myself and the elderly man, each holding fast to the fragile tendrils of our past, our roots entwined with the beloved plants that offer us both solace and strength. Here's one immigrant selling the other a table made entirely out of roots, one that is about to be loaded into a van and travel elsewhere, certainly, very far from where it originally came from.

Chris pulls out. "Goodbye!" I pull up the window and turn to the garage once more. I only catch a glimpse of the door as it lowers to the ground. The warm light slowly retreats onto the cold driveway as a wave retracts into the sea. It's pitch-black

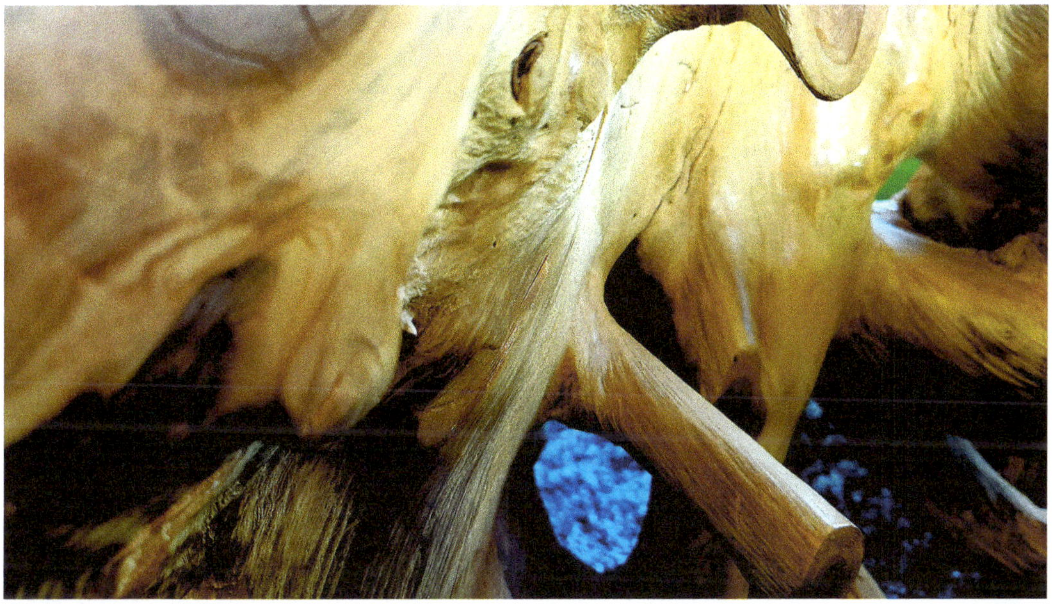

FIGURE 4.2
Giovanni Aloi, *Root Table*, detail, 2023.

FIGURE 4.3
Giovanni Aloi, *Echeveria lilacina*, 2023.

FIGURE 4.4
Giovanni Aloi, *Passion Vine Herbarium Plate with Plants from Author's Garden*, 2014.

again. The van rattles; the noise of the engine fills the air.

As we move away, I imagine myself standing in that garage in the dark, silently surrounded by the lemon trees—the intoxicating fragrance of the bloom. My eyes close. The lemon blooms of Hayat's father smell just like the ones I grew up with in the south of Italy—the same bright and sweet scent. Two distant lands, two men whose encounter is so unlikely, one plant bridging their distant lives, momentarily outlining a trajectory of vegetal love—a bond difficult for words to express but one that plants themselves know how to convey perfectly (Figures 4.2–4.4).

Zimmerlinde: A Refugee Plant

The elongated stems, the heart-shaped leaves, and the vivid lime green—during the 1920s and 1930s, *Sparmannia africana* (also known as zimmerlinde) became a staple in the living rooms and studios of upper-class cultural practitioners across Continental Europe. A versatile and resilient plant, zimmerlinde needs little care, and although it originates from the South African Cape, it thrives in partial shade, a common condition in domestic settings. If kept outdoors in temperate to warm climates, it might bloom several times a year: papery white, rose-like petals stretch backward to reveal a golden and blood-red crown of showy pistils. Indoors, blooming is rare. Next to no written documentation exists that commemorates the zimmerlinde's heyday as the quintessential houseplant of the mid-century Continental European living room. But a glimmer of how special this plant once was can be gathered from clues and fragments left behind by an illustrious family who made zimmerlinde an unofficial emblem: the Freuds. Its story, as might be expected, is wrapped in mystery, and like all mythologies surrounding an iconic figure, multiple and at times contradictory versions complicate a straightforward and coherent recollection of events.

According to some, the story begins in Vienna. Sigmund Freud's studio at Berggasse 19 was located on the first floor of an elegant building in the Alsergrund district. It consisted of several rooms, each serving a specific purpose in Freud's practice. The waiting room was furnished with comfortable chairs and a small table. It was adorned with various objects that reflected Freud's personal interests, such as ancient artifacts and sculptures. Adjacent to the waiting room was Freud's consultation room, where the therapy sessions took place. Books and images played important roles in the interior's aesthetics and mood, as did plants. At least three zimmerlinde grew in the studio in large pots. One stood near the famous couch where Freud's patients reclined for their therapy sessions. Another zimmerlinde was positioned near his desk. And one lived in the library, by the bookshelves.[1]

In March 1938, the annexation of Austria by Nazi Germany, the anti-Semitic policies, and the growing violence against Jews created a climate of fear and uncertainty, prompting Freud to consider leaving his beloved city behind. A few months later, in June, the Gestapo subjected Freud's daughter, Anna, to intense questioning and intimidation, hoping to extract incriminating information about her father's activities and assets. This incident further heightened Freud's concerns for his family's safety and convinced him that it was time to leave Vienna for good. Freud and his family thus embarked on a perilous journey to escape Austria, abandoning the home, possessions, and life they had known for decades.[2]

The escape was a carefully planned and executed feat requiring meticulous and strategic decision-making. One of Freud's most trusted allies was Princess Marie Bonaparte, a French psychoanalyst who played a crucial role in securing the documents needed for the family's safe expatriation. They carefully packed what they could, including an extensive collection of books and antiquities, and arranged to ship them to their intended destination of London, England. Freud's escape involved a perilous journey by train and then by boat. Upon arrival, they settled in a house on Hampstead Heath provided by the British Society for the Protection of Science and Learning.[3]

How likely is it that, amid all this, Freud might have considered packing the zimmerlinde plants? Could they have survived being boxed up with no light or water during many weeks of transit? Or did Freud take cuttings of the plants and store them safely in order to root them upon arrival? The chances that Freud might have taken seeds from those plants are slim since they rarely, if ever, bloom indoors, and even when they do, pollination does not happen easily.

The question of the true origin of Freud's zimmerlinde became central to my research on the work and life of his nephew Lucian Freud, one of the most famous artists of the last century. Lucian Freud, the renowned "master of the modern nude," painted zimmerlinde multiple times throughout his career, and he is mostly responsible for keeping the myth of this family phyto-emblem alive. Annie Freud, the artist's first daughter, told me her version of the story in 2018, when we met to talk about her father's extensive and, at that point, neglected paintings of plants. She confirmed the official version, according to which Sigmund had brought the zimmerlinde to London. But she also suggested that Lucian's father, Ernst Freud, her grandfather and a respected modernist architect, might have traveled to the United Kingdom with his own cuttings. Annie showed me a picture of Ernst's studio, showing the windowsill filled with an impressive lineup of ten potted zimmerlinde.

When Lucian first painted zimmerlinde in 1947, the plant's outline was starkly cast against the white background of a small canvas. The following year, zimmerlinde leaves hovered above the head of Kitty Garman, the artist's first wife, in a pastel work on paper. The plant reappeared again ca. 1950 on a small and newly discovered scrap of canvas and in another undated larger study. Dedicated to the artist's second wife, Caroline Blackwood, *Still Life with Zimmerlinde* (Figure 4.5) is most likely a small fragment of a much larger work that Freud never completed. The attention to detail is outstanding. Freud painstakingly reproduced the finely serrated leaf edges and the minute hairs that cover them. After that, despite its steady and prominent presence in Freud's studio, the zimmerlinde was not painted again until *Large Interior, Paddington* (1968–1969) and never thereafter.[4]

During the summer of 2022, I was deeply immersed in the curation of the *Lucian Freud: Plant Portraits* exhibition for the Garden Museum in London. In conversation after conversation, Christopher Woodward, the museum director, and I became particularly fascinated by the story of the zimmerlinde as a "refugee plant." We spent many meals discussing the alternative storylines, attempting to piece this puzzle together. Just a couple of years prior, the publication of William Feaver's monumental biography *The Lives of Lucian Freud* added a new variation of the story to the myth.[5] The author was Lucian's close friend, and he had obtained much firsthand, and presumably reliable, information from the artist. In the book, Feaver claims that Freud's 1947 zimmerlinde was painted because the plant reminded him of his grandmother's apartment in Berlin.[6] Later in the book is a precious quote, one of the very few times the artist ever spoke directly about his plants: "I've always liked zimmerlinde; my father used to grow them: he used them in winter gardens. The stems

FIGURE 4.5
Lucian Freud, *Still Life with Zimmerlinde*, ca. 1950. Private Collection / Christie's Images © The Lucian Freud Archive. All Rights Reserved 2025 / Bridgeman Images.

were held up by the windowpane. If I moved it, it would collapse. This one had huge leaves and so it took on a life against the light."[7]

But, perhaps more interestingly, Feaver unveiled a previously unknown story according to which the zimmerlinde painted by Lucian Freud are not the direct descendants of Sigmund's or Ernst's plants but rather later acquisitions from his grandfather's family doctor, Professor Hamburger, who also fled Berlin in 1933. The doctor's son, Michael Hamburger—a close friend of Lucian—claims that, in 1947, the artist asked if he could take a cutting from his family's plant to paint it. According to Feaver's account, the Hamburgers' zimmerlinde were allegedly the descendants of a leaf sent to Michael's mother in an envelope from Germany. Apparently, Lucian was given many cuttings.[8]

I remain very skeptical of this story, for I believe that it is impossible to propagate a zimmerlinde from a leaf. The zimmerlinde is not a succulent, and, in any case, the leaf would not have had a chance to reach its destination alive due to the slowness of the mail service at the time. Could they have carried the plants in pots on the train and boat? Woodward conceded that at least part of the story was unreliable at best but remained unconvinced that the plant could have traveled directly from Germany in the hands of either Sigmund or Ernst. That was until an email came to his inbox a few weeks later.

A lady named Anne shared with us her family story, which also poignantly centered around zimmerlinde:

> I wanted you to know that I was hugely impacted by the painting *Still Life with Zimmerlinde* when I came to see the exhibition *Lucian Freud: Plant Portraits*. At first, I didn't recognize the plant from across the room. Later, on close examination, I realized that several pots of this plant used to grow in my parents' house in North-West London when I was a child. Like Sigmund Freud and Professor Hamburger, my father, a dentist from Bavaria, brought with him either a cutting or, more likely, a potted plant when he fled to England from Nazi Germany in the 1930s. This plant—and several generations of cuttings afterward—were a feature of my parents' living room for the next 50 or so years. My family took the plants with them wherever they moved, and they continued to thrive after my father's death. When my mother passed away in 1985, and the family home was sold, I took a cutting with me to my house in South London. Alas, even though I pride myself on having green fingers—like my late father—my zimmerlinde never took. I concluded that the conditions in my then-house must have been unsuitable, so I did not take the risk of trying to replace the plant. Then as now, the zimmerlinde plants are hard to source and propagate and expensive to buy in the UK. Until I saw the painting in your exhibition, I hadn't given much thought to the zimmerlinde for several decades! However, since my visit, I have been reflecting that, though I didn't experience a great deal of tenderness from my late father,

> I have inherited his horticultural sensitivity, which is particularly important to me in my older years. So, thank you for facilitating this insight.
>
> I have a hunch that more zimmerlinde plants and cuttings made their way to the UK with Jewish refugees from Austria and Germany in the 1930s and that there are other Jewish families in the UK who still have stories about these plants.[9]

I encountered Lucian Freud's very own zimmerlinde plant in 2018, still alive and thriving in the garden of his assistant David Dawson. I can't hide that meeting the very plant Freud painted multiple times was truly special. I never met Freud in person. He passed away in 2011, well before I had finished my PhD or written a single book. At that time, he was an art-world star who likely wouldn't have spared a minute to discuss his plant paintings with me, then a graduate student. So meeting his zimmerlinde was, in a sense, the next best thing, somehow even more personal than meeting the artist's daughter or assistant of twenty-five years. I pondered these feelings for some time. How can plants extend someone's life in such a powerful way? How can they summon the absence of a loved one into the present through the silence of their livingness? How can their vegetal beings contain the distilled essence of a previous identity with such gravitas?

Succulent Legacies

Everything has been hauled away quickly. Three hours was all it took for a crew of five burly men to box nearly twenty years of my London life. The flat is empty. Desolate, the shelves are transfixed by the warm afternoon sunshine. It now reaches deep where, until now, it was never allowed to shine. I stand in an empty shell, at once familiar and alien. Somewhat terrifyingly, the walls remind me of when I first walked in, five years earlier. Then, as an excited and trepidatious first-time homeowner, I walked through every empty room, staring at the bare walls and envisioning my future. Today, all I can read upon them is memories. I don't know what their future holds.

Through the patio doors, once more, I gaze at the garden. How many times have I said goodbye to the plants I cannot take with me to Chicago? There's nothing left in the apartment, but there's a whole world outside that I can't bear to let go of.

This was my first garden. The first I grew; I planted it as I wished, without fearing what the landlord might say. There wasn't much when I arrived: a patchy lawn that I promptly shrank to make room for more flowerbeds. An old and twisted buddleia toward the back, in need of TLC, became a wonderful shade provider for the ferns and euphorbias that I would not have been able to grow otherwise. Towering cardoons, oleanders, jasmine, and passion vines. Much of what I grew there was part of my vegetal mythology—iconic plants from a childhood spent in the south of Italy. Plants that have always connected me to the faraway people and places that I still long for today.

I'd love to take them all away with me. But they can't come for practical as well as legal reasons. A couple of months before my departure, I made a herbarium with some of their leaves arranged against white cards. I turned my beloved plants into relics—indexical traces of our kinship, semipermanent echoes of their vegetal significance.

It was while pressing these plants that the idea of taking some with me, after all, started to spring. But how? Succulents! Unlike zimmerlinde, which would not root from a leaf, succulents can. I thought: I'd take cuttings. Put them in a Tupperware with a lightly damp kitchen towel and post it to my Chicago address. I read online that plants can enter the United States as long as they are not rooted in soil.

Three weeks later, the post lady rings the bell at my new Chicago home. I look at the parcel and smile politely. I can't possibly guess what is inside. "I haven't ordered anything," I think. Then, I notice the British stamps; I recognize my own writing.

A peculiar kind of Droste effect!

I suddenly remember . . . I thank the post lady and furtively make my way inside, secretly glad my plan worked well. I open the Tupperware with trepidation. There they are: emerald gems of all shapes. I prepare a bed of fresh sand and soil in a tray, carefully position each leaf, and mist them a bit. I place the tray outside in the warm summer air but away from the direct sun. There I look at them while I drink coffee; I track this odd feeling of wholesomeness for a few minutes until it fades. It's as if a part of my London life had finally joined me. Past and present converge across the vast Atlantic, not in the lifeless stillness of my countless other possessions adrift in some forgotten seabound container, but in a rare and irreplaceable communion, a kind of vegetal transmission that pulses with quiet wonder.

There are quite a few different leaves in the parcel, but the most important by far are those of *Echeveria lilacina* (Figure 4.3). A gorgeous, light gray, almost blue succulent with compact and pointy rosettes of leaves. It is not rare by any stretch of the imagination. Mass-produced for summer planters, it's a staple in gardening centers around the world. But this plant is special to me since it is a descendant of one owned by my grandmother, who died in the late 1990s. In the 1970s, she gave my mum a rooted cutting to take with her as she migrated with my dad to Milan. In turn, decades later, my mum gave me a couple of cuttings to take to London in the 1990s. Through thick and thin, this echeveria has become an emblem of the migration stories that have shaped my family generation after generation. How could I leave all of that behind by interrupting that vegetal bond?

Leaves sprouted roots, then stems turned into branches. Blooms appeared and have returned every year since. Ten years later, the plant is still thriving. It's not the same plant that my grandmother passed down to my mother, nor the one that my mother gave to me. And yet, at the same time, it is. It is the sum of their love for each other and of mine for them both, crystallized in the continuity of living vegetal form. Despite looking similar, a replacement plant of the same species could never occupy the same biographical space. Neither could I just gift a cutting of it to someone whom I do not consider family. Some plants can become living relics—connections to loved ones no longer within reach. They are among the most precious, and yet they can be as cheap as chips at the gardening store.

In the opening of my edited book *Why Look at Plants? The Botanical Emergence in Contemporary Art*,[10] I recalled how the potted plants my grandmother grew in southern Italy in all sorts of random repurposed containers have impacted my relationship with the vegetal world: "pansies, variegated petunias, the sculptural dish-like foliage of leopard plants, red geraniums, deep red gloxinias, tall ferns, fuzzy fern-asparagus, a giant rubber plant, slender orange and yellow zinnias, African daisies, sharp mother-in-law tongues, dark-leaved Irish roses, two small palms, nasturtium, nasturtium, nasturtium, and a bonsai-like money-tree."[11]

These are only some of the plants that today have become staples in my home and garden. Hibiscuses and angel trumpets also occupy a prominent place in a series of vegetal mythologies that bridge faraway places, plants, time, and people—my roots are intertwined with those of the plants that populated my childhood or with others that for one reason or another have become part of my own mythologies along the way. Vegetal mythologies can develop at any time in one's life. They are markers of life nodes: meaningful moments, moments of shared growth, blooming, and seeding. Vegetal mythologies are fictions that give our lives structure in subtle and often unacknowledged ways. Like the mythical tales central to ancient Japanese, Indian, or Chinese cultures, or to classical Greece, these narratives are filled with characters linked to reality and yet charged with a sense of magical enchantment that surpasses nature's laws. Vegetal mythologies magnify the memory of certain plant encounters. They transfix and perpetuate. Science and philosophy have taught us to relegate those instances of affect between us and plants to the realm of romanticized futility. Too personal to be shared with others in meaningful ways outside of poetry or novels, these fictions are still today not taken seriously. But most importantly, vegetal mythologies pry open epistemological spaces between scientific thinking and Western philosophy. This interstice is ripened with an epistemological potential we have been taught to disregard as romantic or irrelevant. It is here, in this space, that we can reclaim ourselves and our firsthand experiences with plants as legitimate and valid sources of important knowledge production.

Vegetal mythologies may refrain from anthropomorphizing the plant, but they personalize the relationship to the point that plants acquire an irreplaceable kind of individuality. Plants can be propagated via seeds, grafting, and cuttings. Among others, these methods allow the same plant to become ubiquitous—a living lineage that connects people across faraway places. Multiple and yet individual, my *Echeveria lilacina* is still with me. In its embodied livingness, diffracted across the branching structure and disseminated through the pointy tips of its blue-gray leaves, is something that no other plant, object, or photograph could contain or express. It is in its vegetal silence that multiple voices, past and present, entwine, indissolubly, seamlessly, unabated.

Uprooting and Grafting: New Vegetal Reconfigurations

Those of us who have left our native countries to settle elsewhere know what it means to feel uprooted. Whether voluntary or not, the separation always entails a complex cultural and emotional negotiation. To not recognize the land, to not fully comprehend the language; to be present and yet thoroughly ungrounded. Plant metaphors such as uprooting and transplanting have been frequently used in the historical contexts of diasporas for good reason. Their recurrence teases to the fore at least part of our latent plant-being—the ancestral planthood we carry within us, but that we have been taught to forget. When our joint evolutionary history with plants began, one and a half billion years ago, the eukaryote division into plants, fungi, and animals entailed a radical shift in the relationship with the ground for some earthlings.[12] Plants and fungi opted to stay close, gripping tight to the earth. But animals chose to flee; some followed their desires with such unwavering resolve that they soared into the sky. And yet, our evolutionary conquest of mammalian freedom is only partially real, for it is underpinned by invisible roots that in one way or another tie us to our lands of origin. This is why the trauma of uprooting is inescapable. Diasporas are often tragic. None are alike. They unfold at different speeds. Some unfold over weeks. Others take decades to manifest. There's a substantial difference between a voluntary transplantation and an uprooting. The chances of a successful transplantation in a new land are often lower when one is uprooted, just as an abruptly eradicated plant will most likely perish while one that has been carefully coaxed out of the ground and repositioned elsewhere might eventually settle and thrive.

The traumatic nature of uprooting experiences often eludes linguistic representation. The overwhelming sense of loss and longing, the loneliness, the vulnerability; the sense of unsteadiness and precarity that unfolds subtly, over time, in waves, underpinning every moment, tinging life with a kind of barely perceptible and yet omnipresent wrong-placeness that strikes when we least expect it: in the mispronunciation of a word, the misplacement of an accent. The endless negotiation between space and time, past and present. Finding ourselves uprooted is always a new beginning. But it is never an easy one. Uprooting inevitably fragments and ruptures. It jeopardizes a wholeness that can never quite be reconstituted on new land. As is the case with many plants, when uprooted, the radical system becomes stunted, for a certain period of time at least. Old relationships are severed as new ones are made. Growth is slowed down as acclimatization to the new soil requires a lengthy process of attunement. Only time will tell if thriving might unfurl.

Uprooting defies the fictitious unity of being, the finitude of the human that was theorized by the humanist philosophies of the Renaissance. Bolstered by Christianity, humanism reaffirmed the supremacy of humanity above all other beings on earth. This anthropocentric philosophical current fueled colonialist practices and the modes of violent uprooting that have defined our modern world. It also configured the philosophical background upon which slavery, genocide, land dispossession, and environmental devastation could be carried out and justified. Climate change might find its chemical origin in the Industrial Revolution, but the climate crisis has its roots firmly planted in the philosophies of the Renaissance. Whether it's the existentialist's confrontation with death, the phenomenologist's exploration of being, the Stoic's acceptance of life's transience, the Kantian limits of human knowledge, or the Nietzschean eternal recurrence—each of these

FIGURE 4.6
Otobong Nkanga, *Diaspore* performance during "14 Rooms," Basel, 2014.

theories falls short of nearing the complexities that truly define the poetic dialectics of the living. Uprooting, in its radical dislocations, reveals the intimate importance of the connection to human/more-than-human networks that often remain invisible, latent in both plants and humans.

It is in this context that art can provide important lines of thought and references to help us grasp the subtlety of our experiences. The work of Otobong Nkanga and Fatma Bucak teases these unspeakable nuances into visibility; they make them flicker momentarily through the media of installation and performance. Both artists have directly experienced uprooting. Nkanga was born in Lagos, Nigeria, and relocated to Paris during her childhood to follow her mother's work. The premature loss of her father meant that seeking financial stability abroad was inevitable.[13] Bucak was born in Turkey and relocated first to Italy and then to the United Kingdom.[14]

Nkanga's performance, *Diaspore*, is deeply rooted in the artist's own personal experiences as a Nigerian-born artist who has lived and worked in various countries (Figure 4.6). Her performances often draw from her cultural heritage as well as the histories and contexts of the places she has lived in or visited. By incorporating elements of her own identity and experiences, Nkanga brings a unique vegetal perspective to the concept of diaspora, highlighting the interconnectedness of different cultures and the impact of migration on individuals, communities, and the land. Nkanga's performances often aim at creating a space for dialogue and reflection, encouraging viewers to consider their understandings of identity as a construct shaped by their cultural backgrounds and relationships with the land.

In this context, *Diaspore* is a multisensory experience that evokes both the physical and emotional dislocation that comes with being part of a diaspora. The performance integrates elements of sound, movement, and visual art to reflect the feeling of "in-betweenness" that many diaspora communities experience. The performer(s) often navigate shifting spaces, mirroring the instability and rootlessness that define diasporic existence. Women of African descent each carry a potted night-blooming jasmine on their heads as they dance over a heavily stylized map that simultaneously echoes those in geographical atlases and dance-step diagrams of the kind that are laid on the floor to learn a new dance. Tracings, borders, lines, bands, quarters—where to stand and what not to cross? The performance offers a glimpse into the endless process of negotiation that immigrants always engage with, oftentimes throughout their lives. Dancing with, against, following, recoiling. Whose toes should one not step on? Who is going to become an ally, or even a partner, in this dance? Stepping into someone else's territory is just as unavoidable as the consequences that will inevitably unravel. Nkanga's performance thus reflects on the blending of cultures, languages, and practices that occur in diasporic events, challenging monolithic ideas of culture and identity and demonstrating the fluidity and complexity of being "in-between" multiple cultural worlds. The body becomes a site for exploration of diasporic subjectivity, embodying both the trauma and resilience of displacement. The movements and gestures performed often evoke the struggle for survival, belonging, and self-definition within the constraints and challenges of diasporic realities.

The title of the performance, *Diaspore*, not only directly references diasporas but also the structure that, in plants, is responsible for the dissemination of seeds. Within certain species, the diaspore refers to the combination of a seed and fruit, or a seed and elaiosome: fleshy compounds of lipids and proteins.[15] These often attract ants and other insects who, drawn by the nutrients, transport the seeds underground to their burrows where they will eventually germinate. Through these vegetal metaphors, Nkanga examines the experiences of displacement, migration, and cultural hybridity that are inherent in the diasporic condition. Through her performances, she delves into the complexities of identity and belonging, highlighting the ways in which individuals navigate multiple and often uncharted cultural, social, and geographical spaces at once.

The plants that Nkanga's performers carry on their heads and shoulders become emblems of the complexities that immigrants experience. *Cestrum nocturnum* is commonly called night-blooming jasmine, but it isn't a jasmine. It is a species of plant in the Solanaceae (nightshade) family rather than the Oleaceae family, to which true jasmines belong. The plant is grown across the globe as an ornamental variety, but it is originally from the West Indies and has today naturalized in Australia, New Zealand, South Africa, southern China, and the southernmost United States.[16] Its adaptable versatility and resilience have made it difficult to keep under control, and, in many countries, it is now considered an invasive weed. Nkanga's performers carry all of this burden and more on their heads and shoulders. The hardship, discrimination, mistrust, and the often relentless exoticization that others under the pretense of admiration. The struggle is visible; the resilience, undeniable.

The fragility of uprooting and the slow process entailed by emigrational assimilation is further exemplified in the precarious livingness of Fatma Bucak's *Damascus Rose* (2016) (Figure 4.7). Sometimes grown in individual pots—emblems of portability and misplacement—and at other times situated in gardens, Bucak's roses tell stories of cultural identity, loss, and regeneration. Damascus roses are not just any ornamental rose. The history of this cherished plant can be traced back to ancient times, particularly in the Middle East and Mediterranean regions. This variety of rose is known for its exquisite fragrance and beauty, and it holds cultural and symbolic importance in Syria. Its name derives from the region around Damascus, where it has been cultivated for centuries to produce rose water, essential oils,

FIGURE 4.7
Fatma Bucak, *Damascus Rose*, 2016. Gardens of damask roses propagated through grafting techniques with rose cuttings from Damascus, Syria. Installation view, *And men turned their faces from there*, David Winton Bell Gallery, Brown University, Providence, R.I. Photograph by Jesse Banks III.

dried blossoms, and preserves, which are utilized in fragrances, cosmetics, and traditional Syrian cuisine.[17]

This versatility has also played a significant role in the country's economy. Its cultivation and oil production have been important industries in Syria for centuries. The rose oil, known as "attar of roses," is highly sought after and has been exported to different parts of the world. The rose industry has also provided employment opportunities and contributed to the country's economic growth. This prominence has been celebrated in literature, poetry, and art throughout history and has become a cultural icon, embodying the essence of Syrian heritage. In addition to its cultural and economic significance, the Damascus rose has also played a role in traditional medicine. Its petals and oil have been used for their therapeutic properties, such as soothing skin conditions, reducing inflammation, and promoting relaxation. The rose's medicinal properties have been passed down through generations and are still used in traditional healing practices. However, over the past few decades, the Damascus rose has come under threat of disappearing because of the civil war that started in 2011.[18]

Bucak draws on her own experience as a member of the Kurdish minority in Turkey as well as the state of the world today, when violence, migration, dispossession, and repression have significantly altered human life. Her grafts of Damascus cuttings on European or American rose plants point to the complexity of integration and assimilation for those who migrate, abandon their roots in their homeland, and adopt new ones in new locales. Grafting is a horticultural practice that involves combining different plant parts to create a new hybrid. This technique has been used for centuries in agriculture and gardening. An incision is made on the surface of a branch or stem in order to expose the pulp underneath. The new cutting is then carefully inserted into the wound so that both ends eventually bind and seamlessly suture. Bucak's grafting of multiple rose varietals onto a single stem reflects the fusion of diverse cultures and the complexities of identity construction in the context of migration. At what point has the original Damascus rose become a different species? The blooms might resemble those of the native plant, but the roots that feed it are no longer Syrian. With its historical and cultural ties, the flower becomes a symbol of the possibility of harmonious blossoming across civilizations. However, the artist is deeply aware that not all attempts at grafting will take. In her work, she embraces failure as part of her process. Some grafts are stronger than others and will bloom; some never will, and others will be rejected by the new mother plant, then shrivel and die. These dimensions are an integral part of the installation.

Bucak's artwork thus prompts viewers to reflect on the fluidity of identity and the ways cultures can influence, enrich, or obstruct one another. It also serves as a reminder of the importance of valuing and safeguarding cultural heritage while embracing the potential for growth and transformation through the exchange of ideas and influences. In the artist's hands, the act of grafting becomes a subversive gesture, one that questions the boundaries and divisions imposed by society, history, and time. It symbolizes the potential for transformation, resistance, and resilience, as well as the power of art to challenge and reshape existing structures.

It is not a coincidence that both Nkanga and Bucak have opted to work with plants to express these complex ideas. In the semantic framework outlined by these artists, the livingness of plants embodies the unstoppable forces of time's passing—the type of process that shapes us into who we are and that is never just down to desires, dreams, and drives. Plant identity, as Michael Marder has conceived it, is inexorably shaped by the environment in which a plant grows, a "sheer exposure in space and time." He continues: "Given that the plant's self, bound to the universality of the elements and of light, is always external to itself, its unity is at once a disunity."[19] It is this very conception of vegetal disunity, one that nonetheless can exist in an apparently circumscribed, autonomous body, that can help us appreciate the subtleties of certain circumstances that we experience on our skin and bones and that the outdated humanist conception of finitude has obfuscated and concealed. It is to this heightened awareness that plant-being provides important metaphors that might help us to better comprehend and grasp the nuances involved in diasporas of all kinds—metaphors that might make the struggle bearable and perhaps even help the graft to successfully take.

Notes

1 Conversation between Giovanni Aloi and Annie Freud, the great-granddaughter of Sigmund Freud, June 2018.
2 David Cohen, *The Escape of Sigmund Freud* (London: JR Books, 2009).
3 Cohen, *The Escape of Sigmund Freud.*
4 For more on Lucian Freud's paintings of zimmerlinde, see Giovanni Aloi, *Lucian Freud Herbarium* (New York: Prestel, 2019); and Giovanni Aloi, *Lucian Freud: Plant Portraits* (London: The Garden Museum, 2022).
5 William Feaver, *The Lives of Lucian Freud: Youth, 1922–1968* (New York: Knopf, 2019); and William Feaver, *The Lives of Lucian Freud: Fame, 1968–2011* (New York: Knopf, 2021).
6 Feaver, *The Lives of Lucian Freud: Youth,* 437.
7 Feaver, *The Lives of Lucian Freud: Youth,* 1011.
8 Feaver, *The Lives of Lucian Freud: Youth,* 437.
9 Email, March 3, 2023.
10 Giovanni Aloi, ed., *Why Look at Plants? The Botanical Emergence in Contemporary Art* (Leiden: Brill, 2019).
11 Aloi, *Why Look at Plants?,* 1.
12 New Scientist, *How Evolution Explains Everything About Life* (London: John Murray, 2017).
13 Yvette Mutumba, "Otobong Nkanga: Nothing Is Like It Seems, Everything Is Evidence," *Afterall: A Journal of Art, Context and Enquiry* 37 (2014): 52–59, https://doi.org/10.1086/679376.
14 Jess Chen, "Fatma Bucak," *Art Forum* 61, no. 2 (2022).
15 Pierre-Michel Forget, Joanna E. Lambert, Phil Hulme, and Stephen B. Vander Wall, *Seed Fate: Predation, Dispersal, and Seedling Establishment* (Wallingford: CABI, 2005), 180.
16 Quentin C. B. Cronk and Janice L. Fuller, *Plant Invaders: The Threat to Natural Ecosystems* (London: Chapman & Hall, 1995), 146.
17 Joseph H. Pemberton, *Roses: Their History, Development and Cultivation* (London: Longman, Green, 1908), 66.
18 "Withered by War: Syria's Damask Rose Is Under Threat," *The National,* May 17, 2016, https://www.thenationalnews.com/world/withered-by-war-syrias-damask-rose-is-under-threat-1.203865.
19 Michael Marder, *Plant-Thinking: A Philosophy of Vegetal Life* (New York: Columbia University Press, 2013), 89.

EMPLACEMENTS

5

"JOYOUS HOWLING"

Hoolock Gibbons, Tea, and Plantation Landscapes

ROMITA RAY

A hoolock gibbon stands on its toes in a Company painting, with one hand grasping a branch and the other dangling below its knee, its fingers curled upward (Figure 5.1).[1] With its long, floppy arms (seemingly) pinned to its torso just below its chin and its eyes hollowed out in a mask-like face tilted upward, the animal is more puppet than primate. The only nonhuman ape species to be found in India, this arboreal creature lives high up in *Oolung* (Hoolong) and Mackoi trees, where its "joyous howling" rings across the forest to join the calls of fellow gibbons (Figure 5.2).[2] Yet no traces of its vibrant life world nor of its canopy habitat, where it swings with "astonishing rapidity from branch to branch," linger in the image. Instead, we encounter a stiff figure reassembled from animal skin and displayed on a leafless branch—the latter a standard prop in scientific illustrations and taxidermy exhibits of the time.[3] Sentience has given way to the confluence of field research and the visual arts in the practice of science that now served the economic incentives of the colonial state.[4]

Collected in a remote corner of northeast India, the hoolock is a harbinger of the radical changes that would sweep across its habitat while political tensions unfolded thousands of miles away, between Calcutta (Kolkata), Canton (Guangzhou), and London. In 1835, a few years after the East India Company had lost its monopoly over its China trade, the government of India dispatched a team of naturalists to "visit and inspect the Tea-forests" of Assam with the goal of verifying a *jungli* (wild) variety of tea as a *Camellia sinensis* species.[5] This fact-finding mission would lay the foundation for commoditizing tea in India and, in the process, help diminish Britain's dependence on Chinese imports.[6] Reinvesting British capital in Assam instead of in a weakened China trade was a strategic means of transferring economic "value" from India back to Britain.[7] But where and how to cultivate tea in dense

FIGURE 5.1

Unknown Indian artist, *Simia Hylobates hoolock, Harlan*, watercolor, gouache, pen, and ink on paper, 1836. From the archive of the British Library, NHD6/984.

FIGURE 5.2

Western hoolock gibbon (*Hylobates / Hoolock hoolock*) male hanging from branch. Endangered. Gibbon Wildlife Sanctuary, Assam, India. Nature Picture Library / Alamy.

jungle teeming with wildlife? Economic agendas forged in London and Calcutta were not so easily pursued in "that almost unexplored district" of Assam, where *jungli* tea specimens had been circulating since the 1820s, when British officials, military men, and merchants stationed in the area first exchanged tea samples; two such rare specimens are housed today at Kew (Figure 5.3).[8] What followed, then, was a government-sanctioned survey of the environmental conditions in which a forest species of tea flourished, leading to copious field notes, letters, and specimens that came to anchor the search for tea in the rhythms, pauses, and flows of the jungle. As concrete glimpses of the "extensive forests" of Assam crystallized, botanical and animal fragments hovered at the capacious interface of collecting, observing, and translating nature into science, their physicality disclosing the ways and means through which the detritus of an "artful science" (borrowing from Barbara Maria Stafford) materialized in the nineteenth century.[9]

Seen within these broader vegetal and animal epistemologies, the Kew specimens suddenly look very different. Juxtaposed with the mammals and birds collected during the tea expedition, they gesture at the collective web of plants and nonhuman animals that shaped the vitality of the jungle into whose dynamic matrix members of the tea delegation inserted themselves for several months, navigating its challenging terrain and recording its geography and natural history.[10] It is to these broader entanglements that the hoolock belongs, its image in the Company painting offering us a fragmentary glimpse of the "*zoology* of the province" whose "careful examination," according to geologist and expedition member John McClelland, was second only "to the relations of the [tea] plant in regard to soils, and its association with other vegetable productions."[11]

Soil, plants, and animals constituted tea-growing terroir. Understanding their manifold relationships was key to confirming the potential of cultivating tea in Assam as well as to comparing the area with the "tea districts of China."[12] But the area's dense tangle of flora and fauna made it difficult to observe, let alone to record, the composition of tea-friendly terrain. After all, "impenetrable tree forest" and "grass jungle" defied what Bikash Sarma describes as "the logic of visualization."[13] Collecting plants and animals was one way of coming to terms with the opacity of *jungli* foliage. Specimens reduced the density of the forest to a manageable scale while offering evidence of a diverse ecosystem that remained a geographical mystery. They were also imprints of the shapes and textures of the jungle from which they were extracted. Quite simply, specimens constructed, spatialized, and visualized the wildness of the wild. And what colorful representations they made, their delicate, painted contours leaping off blank sheets of paper to delight the eye and gratify one's curiosity, their images directing attention to the aesthetic pleasures and not just the empirical pursuits of colonial science. Wildness merged with wonder in a growing corpus of scientific knowledge.

Fast forward now to a photograph taken by the tea planter Frank Nicholls more than a century later, of a hoolock pulling the tail of a pet cat in the bungalow garden of Borkatonee Tea Estate in Assam (Figure 5.4). Unlike the awkwardly posed primate in the Company painting, the animal captured on camera is deeply responsive to its fellow beings. An adept "ball boy" on the tennis court, where it "collect[ed] the balls and return[ed] them to the server when necessary," the primate was (to borrow from John Berger), a "creature" of its owner's "way of life," its human-like behavior suggesting that, like the *jungli* tea plant transformed into a commodity, it too was tamed by the disciplining forces of plant capitalism.[14] By the time the hoolock was photographed, vast stretches of "wastelands, including forests," were cleared out to grow tea.[15] One-fifth of Assam's geographical area was harnessed for commercial cultivation, while more than "4,000 square miles of forest were enclosed and brought under the control of the colonial forest department."[16] A vivid sign of the radical landscaping measures that would continue to alter jungle into plantation well into the twentieth century, the verdant lawn bordered with ornamental plants surrounding the primate affirms an aesthetic of discipline and order ushered in by commercial cultivation. A small locus of intimacy in a sprawling plantation, it registers the flux and flow of a planter's life: family gatherings, parties, interactions with household servants

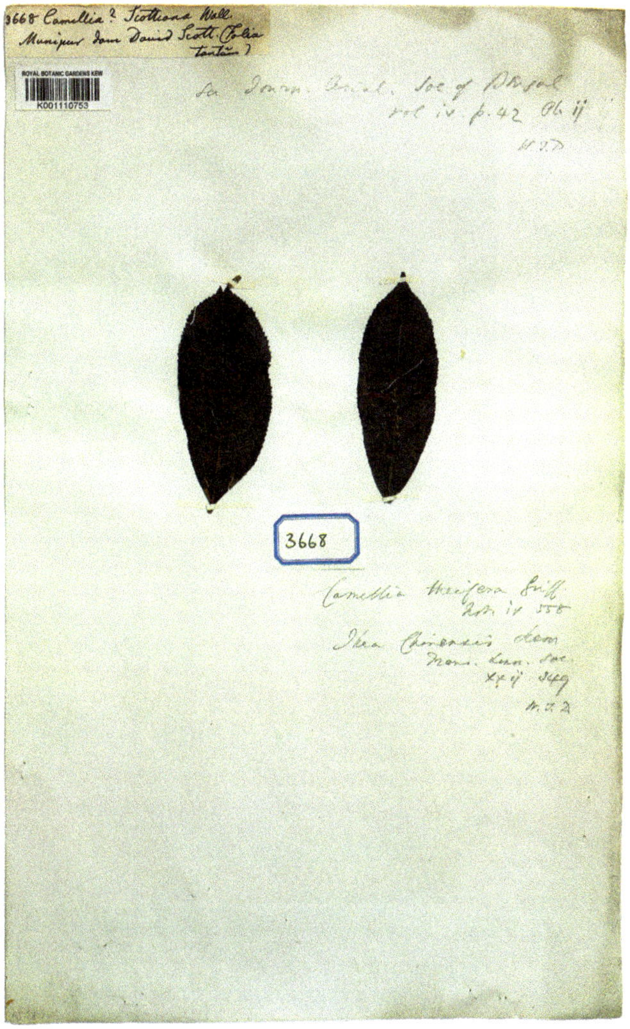

FIGURE 5.3
Camellia scottiana Wall, [family THEACEAE]. K001110753, Royal Botanic Gardens, Kew. © The Board of Trustees of the RBG, Kew.

FIGURE 5.4 "The playful hoolock teasing the cat. He was also a 'ball boy' on tennis days," from Frank Nicholls, *Assam Shikari: A Tea Planter's Story of Hunting and High Adventure in the Jungles of North East India* (Auckland: Tonson Publishing House, 1970), fig. 46.

and gardeners, and, as Nicholls's photograph suggests, encounters with pets. Inserted into its grassy enclosure, the hoolock is a survivor, a poignant remnant of the scattered trails of dislocated and disposed plants and animals left behind by wide-scale deforestation.

Dislocation did not necessarily terminate the *jungli*; rather, it opened new—albeit fragile—avenues for the *jungli* to be relocated, rearranged, and reconstituted *within* both tea-growing enclaves and the metropolitan centers of the British Empire. New interspecies relationships and biotic assemblages emerged from its fraught mobilities, the plantation itself being one such constellation of plant-human-animal linkages. As Maan Barua points out, "colonialism and capitalism are ontological projects that impose their own version of the real while violently erasing other ways of apprehending and inhabiting the world."[17] What visual, material, corporeal, and spatial realities, then, do the hoolocks in the painting and photograph represent? What new plant-human-animal alliances and erasures do they bring to the fore in a geography transformed from jungle into plantation, where the forest still held sway over monoculture landscapes? Can we read plants and plantations through *their* bodies? Threading their way through this essay, these questions follow the two primates across space and time as sites of colonial encounters rooted in the *intimacy* of vision and touch, whose vectors consistently underscored the need to reconfigure alien bodies and their physical contexts as familiar entities: specimens, drawings, pets, and gardens. Above all, this intimacy hinges on the tea plant, whose value as a plant commodity inspired the historic tea expedition that yielded the hoolock drawing and paved the way for plantations where *jungli* animals sought refuge or, like the hoolock, were relocated through human intervention.

Sophie Chao reminds us that plantations are not just capitalist "sites of extraction and extinction," where human mastery transforms plants into lucrative commodities at the cost of vegetal and nonhuman animal lives.[18] Nor are they strictly plant-oriented "modular landscapes," as the term "plantation" suggests, from where "unruly forms of more-than-human sociality" are "banished," as

Anna Lowenhaupt Tsing, Andrew S. Mathews, and Nils Bubandt claim.[19] Plantations are fundamentally multispecies habitats that remain knitted together with the ebb and flow of the *jungli*, bringing into sharp focus the entanglements and disentanglements between plants, humans, and nonhuman animals mobilized by commercial cultivation and capitalist systems of wealth production.[20] As the emerging discipline of the plant humanities engages with how plants both shape and are a part of human life worlds, can it make room for the animal interlocutors with whom the material and affective dynamics of plant and human lives remain enmeshed?

I now turn to the hoolocks to investigate this question: first by examining the primate as an object of scientific scrutiny tied to the search for tea; and next by considering how tea-growing enclaves reshaped the animal's social and spatial realities. At the heart of my analysis lies the jungle, whose shifting ecologies remind us that specimens removed from its expanse foreshadowed its own fragmentation and removal to make room for plantations. In the increasingly hybrid terrain of forest and commercial tea cultivation arose a modernity built on the extractive dynamics of the Plantationocene that forever altered forest ecosystems, transformed land into plantation, fragmented communities into labor, and commoditized plants, all in tandem with science and technology that created new networks of connectivity to maximize modes of productivity. What made this emergent modernity unique was its intersection with the *jungli*, the very antithesis of the modern, whose dynamics frequently disrupted and occasionally even destroyed the flows and rhythms of plantation lives. To turn to them is to take stock of how the wild and untamed actively shaped plant capitalism, drawing the dynamics of tea, a plant commodity, into the wider web of interspecies relations that constantly shifted and changed with the ever-expanding footprint of plantation landscapes.

Jungli Fragments

In the nineteenth century, plant and animal fragments produced a vast, albeit uneven, archive of knowledge that thickened and intensified with colonial travel and expansion. Reconfigured as a specimen and watercolor, the hoolock in the Company painting represents this dynamic matrix of science, art, and craftsmanship from which its visual and material representations emerged. Scraped of its flesh and bone and reduced to a skin, the primate was first transported to Calcutta, where, along with other animal fragments collected during the tea expedition, it was recorded by an Indian artist employed by the East India Company. Shipped to London together with its drawing and a "Descriptive Catalog" of all the Assam specimens, the hoolock fragment was deposited at the East India Company Museum, where, "mounted" and displayed, it became "a valuable addition" to the museum's natural history collections.[21] As William Tronzo observes, fragments are not just a "static part of some once-whole thing" but are themselves "something in motion."[22] Chemicals, skin, and fur fashioned new ways of experiencing the *jungli* in Calcutta and London, where, in the tomblike chambers of natural history museums and storage cabinets, a semblance of its body was resurrected, first for scientists and then for an inquisitive public.

Like the hoolock, the tea plant was also taken apart, but its fragmentation into leafy specimens, flowers, leaves, fruits, and seeds was aimed specifically at investigating its potential as a plant commodity, not just as an artifact of science. Circulated in the decade preceding the tea expedition, botanical fragments of tea would shape key exchanges that led to endorsing a wild plant as a genuine *Camellia* species, a flowering plant genus in the family Theaceae—an exciting discovery that would mobilize a relentless plant capitalism centered on what Victorian tea planter George M. Barker would call a "cheering beverage."[23] Tribal chiefs, plant collectors, middlemen, metropolitan scientists, shipbuilders, military men, and colonial administrators, to varying degrees and with different levels of agency, constituted the early networks of actors and agents who would determine the fate of this *jungli* tea. It is from their overlaps and intersections that the hoolock specimen and its images emerged.

In the 1820s, Captain David Scott, British agent to the Northeastern Frontier, sent tea leaves to James Kyd, master shipbuilder in Calcutta (see Figure 5.3). "The enclosed leaves," he wrote, "are said to be those of the wild tea plant. I have not been able to get the flowers or seeds but I have some plants that I hope will [preserve?] & plenty more are procurable. The . . . Baruahs & a China man that is here say that it is the tea plant perhaps the species that Mr. Gardner sent down from Nypal [*sic*]."[24] Kyd subsequently dispatched the leafy fragments to Nathanial Wallich, the Danish superintendent of the Calcutta botanic garden, who rejected the possibility that they could be tea specimens. Nearly a decade later, a government-appointed Tea Committee, for which Wallich would serve as secretary, would acknowledge Scott's contribution, but with the caveat that they had withheld judgment about their authenticity until they had procured the "fruit of the reputed [tea] shrub."[25] Marred by doubt and inexperience, these early years of investigating tea relied on varying levels of tea expertise. An outlier in Calcutta compared to his colleagues in the northeast, Wallich had no direct knowledge of wild tea growing in situ; nor did he have contact with the Indigenous peoples—the forest-dwelling Singhpos and Khamtis—who had consumed tea over countless generations. In 1816, he had similarly (and erroneously) rejected tea samples sent by Edward Gardner, an honorary British resident in the royal Nepalese court in Kathmandu (notably, Scott's letter mentions comparing the tea sample he had sent to Wallich with Gardner's tea plants).[26]

Procuring tea specimens from Assam began in earnest with Scott's revenue collector Maniram Dewan, who, in the 1820s, initiated a meeting between the Singhpo chief Beesa Gaum and the Scottish merchant-explorer Robert Bruce; the latter had come to learn of wild tea plants growing on the Assam–Burma border. An agreement was reached to furnish Bruce with wild tea plants and seeds but was interrupted soon afterward by the Anglo-Burmese War. The transfer finally took place after Bruce's death in 1824, when the Singhpo chief honored the arrangement and sent plants and seeds to Robert's brother Charles, a midshipman in the East India Company's service. Charles, in turn, deposited these botanical fragments with David Scott, who planted some in his garden, sent a few samples to Captain Francis Jenkins, the government officer and amateur botanist stationed in Guwahati, and dispatched the rest to Wallich (via Kyd) in Calcutta, who questioned their botanical pedigree.[27]

A crucial turning point in tea-related exchanges took place in the wake of the First Anglo-Burmese War (1824–1826), when the East India Company annexed hundreds of acres of forested land in Assam, Manipur, and Cachar. Growing hostilities between Britain and China reanimated discussions about the commercial potential of growing tea in the area. A newly minted governor-general, William Bentinck, an ardent Utilitarian who advocated for free trade and capital investment in India, revisited the prospect of tea cultivation by laying out a plan before his council and forming a Tea Committee.[28] *Jungli* tea plants moved back into the spotlight. In response to the committee's call for information that might support the prospect of cultivating tea in British India, wild tea leaves, along with fruits, blossoms, and edible leaves prepared by tribal tea drinkers, were submitted by Captain Francis Jenkins with the help of Lieutenant Andrew Charlton of the Assam Light Infantry. A skeptical Wallich finally endorsed the Assam plant as an authentic species of tea.[29] Soon after this latest reidentification, Wallich would be approached to lead the tea delegation to Assam, where the hoolock was found.[30]

For all the enthusiasm generated by the "discovery" of a commercially viable species of indigenous tea, the Assam plant was widely regarded as primitive, inferior, and a risky investment compared to the China variety that European consumers were accustomed to drinking: "A wild plant is not likely to give as good produce as one that has been cultivated for centuries."[31] But Assam clearly offered tea-friendly terroir, prompting the Tea Committee to settle on Chinese tea plants and seeds to begin tea production in the area.[32] They also favored Chinese tea growers, for as Jayeeta Sharma observes, the "primitive population" of Assam "seemed particularly unsuited to cultivate the delicate tea plant acquired from China."[33] If the politics of race were mapped onto the bio-economics of the plant, then the hoolock was inseparable from these interventions, its wildness amplifying the *jungli* origins of Assam tea that drove the need to produce an "improved" plant commodity.[34] Removing the jungle was a crucial step toward refining the plant.

Reimagining the Jungle

In 1834, the same year that the Tea Committee was convened, the Philadelphia-based naturalist Richard Harlan brought the hoolock gibbon to the attention of the scientific community in America and Europe. Drawing upon Georges-Louis Leclerc, Comte de Buffon's magisterial *Histoire naturelle, générale et particulière, avec la description du Cabinet du Roi* (1749–1804), together with specimens and a "descriptive letter" from Marmaduke Burrough, the American consul in Calcutta, Harlan became the first scientist to identify Buffon's "Orang" as a hoolock.[35] The discovery caused a stir on both sides of the Atlantic and was widely cited throughout the nineteenth century. Based on three hoolocks found in the Garo Hills (now in Meghalaya), Burrough's specimens were procured during his visit to the area: two were living male primates gifted by Captain Alexander Davidson, an East India Company officer stationed in Goalpara, Assam, and another, a living female, was given by a local "Garrow [Garo] Indian."[36] Notably, Davidson assisted David Scott when the latter served as magistrate of Rangpur and as civil commissioner for the districts of Goalpara and Garo Hills, before taking charge, in 1826, as senior commissioner of Assam.[37] It was during their tenure that Burrough visited the northeast and Scott sent wild tea samples to Kyd.[38] By the time Harlan's article was published, Davidson was in charge of Goalpara and Captain Francis Jenkins was appointed commissioner of Assam and northeast Rangpur.[39] Burrough's hoolocks were embedded in the same networks through which the tea specimens at Kew would circulate.[40]

Cut off from their arboreal habitats and communities, all three apes died prematurely.[41] But lost lives made useful specimens and drawings. New materialities and visual logics emerged to participate in and shape the emergent streams of colonial

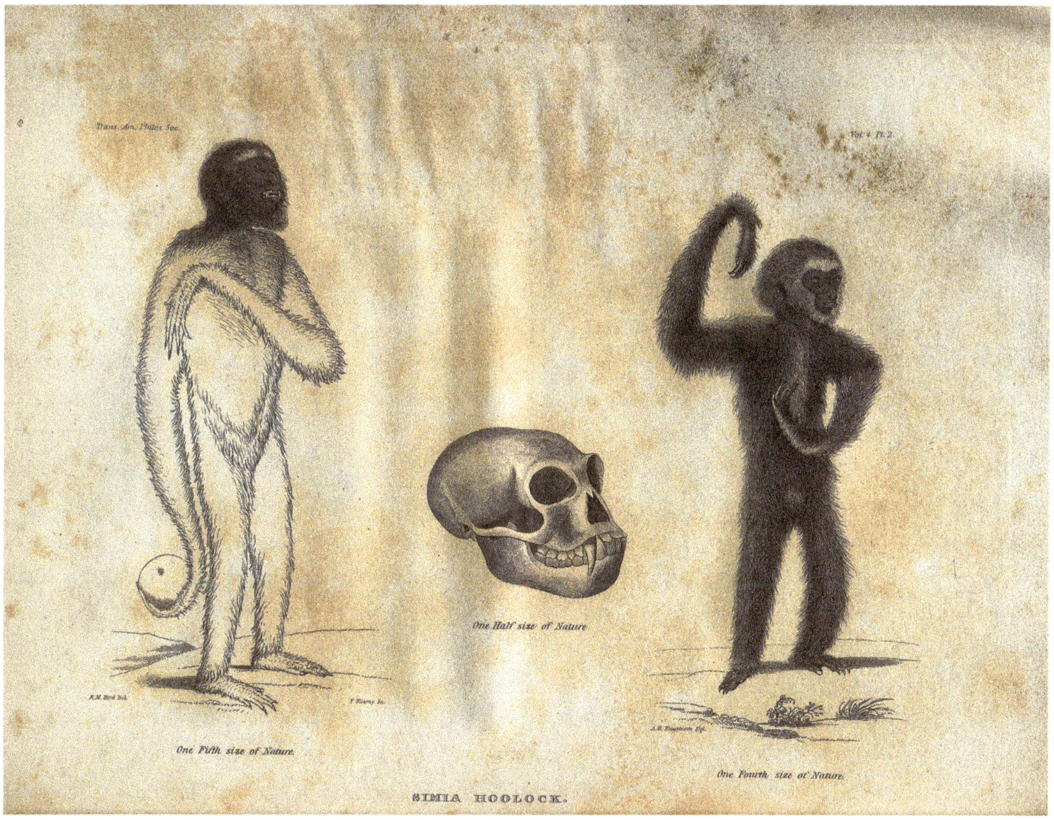

FIGURE 5.5
Francis Kearney, after R. M. Bird and A. B. Engstrom, "Simia Hoolock," from Richard Harlan, "Description of a Species of Orang, from the North-Eastern Province of British East India, Lately the Kingdom of Assam," *Transactions of the American Philosophical Society* 4 (1834): pl. 2. Archives and Special Collections, University of Pittsburgh Library System.

FIGURE 5.6
William Home Lizars, after James Hope Stewart, "Hylobates Hoolock (The Hoolock)," from Sir William Jardine, *The Naturalist's Library: Mammalia; Monkeys*, vol. 27 (London: Chatto and Windus, 1833; reprint, Edinburgh: W. H. Lizars, 1844), pl. 3, hand-colored steel engraving. Album/Alamy.

science that were developing between India, Britain, Europe, and North America. Reconfigured as erect figures in Harlan's article, the hoolocks can be seen on either side of a skull detached from one of their bodies—their long appendages holding a fruit and bending like a dancer's arms (Figure 5.5).[42] The production of science was invested not only in how nature was systemized but also in how nature *performed*. Drawing and taxidermy offered the tools to record and reimagine these patterns of sentience, animating corporeal fragments to explicate the repertoire of more-than-human animal behavior. But animal habits, instincts, and responses to stimuli are mediated by their natural environment. For an arboreal mammal like the hoolock, the spatial thresholds of its *jungli* habitat are crucial to its behavior patterns.

Captured in the "Woods & Jungles" near Goalpara, Harlan's hoolocks were extricated from a "scarce penetrable" wilderness that rendered Cartesian coordinates—so integral to human placemaking—irrevocably ineffective.[43] In these remote areas awash with dense foliage and teeming with wildlife, hoolocks were notoriously difficult to spot, "swing[ing] with such astonishing rapidity from branch to branch, and from tree to tree, that they are soon lost in the jungle or forest."[44] Stretching their long arms to grasp at bark, twigs, and branches as they leapt from one tree to another, they became part animal, part vegetal beings, their appendages turning branch-like as they made contact with the surrounding plants to merge with forest foliage. What gave them away in the complex architecture of the jungle were their piercing shrieks of *hoo, hoo, hooos*—a sonic assertion of their sentience that reinforces what Yota Batsaki calls "the magic of invisibility."[45] "The voice is very powerful and may be heard over long distances."[46]

Yet *jungli* foliage is nowhere to be seen in the illustration accompanying Harlan's article, nor is it present in the Company painting that only hints at the primate's habitat (see Figure 5.1). As in the herbarium sheet to which the tea specimens from Kew are attached, a blank background in the latter stares back at us, its emptiness evocative of what Beth Fowkes Tobin analyzes as the erasure of "local habitat, plant physiology, human use, and cultural significance" in the production of new archival knowledge about flora and fauna.[47] Can the blankness of the page be read as a metonymic sign of deforestation? K. Sivaramakrishnan analyzes the removal of jungle cover in the colonial era as "axiomatic of the principal economic and ecological changes" that placed commercially useful plants front and center in the management of forested land.[48] Clearing out the jungle to make room for monocultures necessitated taking out both unwanted plants and nonhuman animals to establish clear-cut boundaries of property and ownership. And like the tea plant that came to be embedded in these emptied landscapes, the hoolock was transplanted onto a blank page.

In the absence of sufficient raw data, an unknown Assam jungle posed the problem of representation. Depicting its complex morphologies was a difficult, if not impossible, visual challenge. Where information fell short, the imagination took over, as we see in an illustration of the hoolock published in Sir William Jardine's celebrated *The Naturalist's Library* (1833) (Figure 5.6).[49] Here, a leafy overstory invites us *into* the primate's deep entanglements with the vegetal geographies that lie far beyond human reach, its foliage comprising familiar British plants: the sturdy oak, a conspicuous sign of Britain's primeval forests, from which the ape dangles (Figure 5.7); and the fern, another quintessential piece of the wilderness that engulfs its hanging figure (Figure 5.8).[50] As some of the oldest plant species, those "living fossils that embody deep geological time," in Batsaki's words, ferns in particular accentuate the hoolock's primality.[51] They also shaped modern-day spaces of exploration and discovery: from the countryside where naturalists wandered in search of their leafy forms to the "pathless rocks . . . beds of matted fern, and tangled thickets" where poets like William Wordsworth reminisced about walking in the woods; and from the "sealed" chamber of a "glass" container that Nathaniel Bagshaw Ward used to ship "selection of ferns, mosses, and grasses" from "London to Sydney, Australia," to the forests of northeast India where a year later, the naturalist William Griffith, a member of the tea expedition, would encounter "ferns and mosses" near Cherrapunji.[52]

A primal creature turned into a modern specimen, the hoolock's colorful figure imbues its image with a gorgeous immediacy that hints at the plant-animal intimacies of its natural environment. As Felix Driver observes, color enlivens the process of marking and "identifying" a living "presence," while heightening the pleasure of looking, observing, and feeling with one's eyes.[53] Discovery is deeply sensorial. Entwined with the foliage it grasps, grazes against, and climbs, Jardine's hoolock comes across as the quintessential *jungli* creature, a secret of the forest exposed in its (imagined) canopy habitat. All this would change within

FIGURE 5.7
James Ward, *Landscape with Deer Under "The Beggar's Oak," Dagot's Park*, 1820–1830, oil on panel. Yale Center for British Art, Paul Mellon Collection, B2014.5.7.

a few decades when its leafy surroundings would be replaced by plantations.

Jungli Pets and Plantation Modernities

Returning now to the tame hoolock in the photograph by Frank Nicholls, the primate is seen to make physical contact with a cat, another domesticated animal, instead of a plant (see Figure 5.4). Touch, the driving force behind its embeddedness in its arboreal habitat, is now anchored in the fictive kinship patterns that play out between two pets (and their human onlookers) in a planter's garden. Enhancing the visibility of their exchange is the flat, uninterrupted lawn that offers an open field of vision, where the hoolock, no longer obscured by *jungli* trees, is revealed to the human eye. Planted with seed and bordered with dahlias, snapdragons, marigolds, roses, and other ornamental plants most likely procured from nurseries in Calcutta, the garden has replaced the dense jungle where the primate was once a part of a large hoolock community. Isolated from its vibrant network of familial relations, the animal appears solitary and childlike, its smallness accentuated by the camera's elevated vantage point, its severance from its canopy habitat driven home by its proximity to the ground.[54] On plantations, multispecies engagements are allowed to take place, but only on a human scale and within the limits of human sight and touch, the planter's private garden epitomizing one such space where these limitations are tested, reiterated, reinforced, and complicated by the divisions of race and class. Here, flowering plants, human actors and agents, pets, furniture, household objects and tools, and the occasional wildlife intruder cross paths to shape the unique ontologies of plantation life. As Swati Chattopadhyay points out, small spaces and small things "foster connections in unexpected ways."[55]

If the photograph offers conspicuous evidence of the extent to which the *jungli* could be resettled on plantations, then the pet hoolock represents the economic and social realities of what Donna Haraway refers to as "a story of biopower and biosociality, as well as of technoscience" in a highly manicured landscape.[56] By the time the primate made its way into Nicholls's memoir, tea cultivation had

FIGURE 5.8

James Sowerby, "Male Polypody Fern, Polypodium filix mas.," in William Woodville, *Medical Botany: Containing Systematic and General Descriptions, with Plates of All the Medicinal Plants, Comprehended in the Catalogues of the Materia Medica, as Published by the Royal Colleges of Physicians of London, Edinburgh, and Dublin*, vol. 4, 3rd ed. (London: Bohn, 1832), pl. 267. Courtesy of the Missouri Botanical Garden Library via the Biodiversity Heritage Library.

swept across Assam, raising the need for greater connectivity between far-flung, forested environments and the urban centers of the tea trade. As the technological apparatus of telecommunications systems sprang up along the outskirts of plantations and along roads and forest boundaries, the spatial reach of plant capitalism extended far beyond tea-growing enclaves. Telegraph and telephone poles and wires announced "time as [a] territorial medium," replacing leafy trees to mark modern lines of control across an increasingly mixed terrain of forest and agricultural land.[57] Signs of industrial development that encompassed, in Pierre Bélanger's words, "the design of new time scales and new time zones . . . instrumental to the orchestration of large scale effects," these mechanical interventions facilitated the emerging information networks through which the tea industry now operated.[58] Caught up in their spatial systems and strategies, hoolocks would, occasionally, collapse *jungli* and plantation time:

> There was a *telephone wire* which led to another estate of the [tea] company, some two miles away, where there was a local market on Saturday. The [tame] hoolock would see some of the servants leaving the bungalow compound with their usual baskets and away the ape would go. *He would climb the telephone post and then proceed hand over hand for the full two miles, with his legs dangling.* He had many friends at the bazaar—men, women, and especially children. They looked forward to his weekly visits and loaded him with fruit, such as ground nuts and bananas . . . *He returned to the bungalow on the telephone wire in the evening.*[59]

A living site of biocultural effects and interspecies encounters, the pet primate from nearby Sangsua Tea Estate adapts its arboreal locomotion to a telecommunications system. Tea monocultures had ushered in the Deleuzian–Guattarian logic that "complex constellations of objects, bodies, expressions, qualities, and territories . . . come together for varying periods of time to ideally create *new ways of functioning*."[60]

Coping with human-made environments meant yielding to regimens of human behavior. For instance, Harlan (by way of Burroughs) reminisced about an "easily tamed" hoolock that walked hand-in-hand with Burroughs, enjoyed "coffee, chocolate, milk, tea," and responded with genuine affection.[61] Nearly fifty years prior, acclaimed German painter Johann Zoffany had painted a pet hoolock clambering onto the artist's chair in a conversation piece set in Lucknow.[62] But the primate's instinct to roam free disclosed an animality that would continue to linger on plantations. From pets that "[ran] loose about the compound" of tea estates in Cachar to the telephone-wire-traveling primate in Assam, tame hoolocks routinely asserted their agency, serving as powerful reminders that more-than-human animals shaped their own social and physical geographies, as well as their own "bio-geo-graphy" in tea country.[63]

Today, while tea thrives as a commercial plant in Assam, hoolock gibbons are sequestered in the Hoollongapar Gibbon Sanctuary and the even smaller Borajan Wildlife Sanctuary (see Figure 5.2).[64] They can also be found, albeit in shrinking numbers, in nearby Tripura, Manipur, Meghalaya, Nagaland, and Mizoram, and beyond Indian borders in the Chittagong Hill Tracts (Bangladesh) and Myanmar.[65] With sightings on tea plantations becoming rarities, the elusive primate has once again retreated into the imagination. Is there a future for the hoolock in tea country? If so, what might such an ecological futurity look like? These are difficult questions in landscapes that continue to be reengineered by plant capitalism, human settlement, and deforestation, leaving little room to imagine a different yet viable future for an endangered species. If tea determines the fate of the *jungli*, it is uncertain if it can ensure the survival of the primate, or, at the very least, the hoolock's simultaneity in human life worlds.

Notes

1 *Company painting* refers to a watercolor picture made by an Indian artist for East India Company officials, their wives, and affiliates. Distinguished by a hybrid mixture of Indo-Persianate and European painting traditions, these images focused mainly on the flora and fauna of the Indian subcontinent.

2 Charles McCann, "Notes on the Colouration and Habits of the White-Browed Gibbon, or Hoolock (*Hylobates Hoolock* Harl.)," *Journal of the Bombay Natural History Society* 36, no. 2 (1933): 403; and William Jardine, *The Naturalist's Library: Mammalia; Monkeys*, vol. 27 (London: Chatto and Windus, 1833), 136.

3 Jardine, *The Naturalist's Library*, 27:136; and Richard Harlan, "Description of a Species of Orang, from the North-Eastern Province of British East India, Lately the Kingdom of Assam," *Transactions of the American Philosophical Society* 4 (1834): 57–58.

4 Deepak Kumar, "Science and Society in Colonial India: Exploring an Agenda," *Social Scientist* 28, nos. 5–6 (May–June 2000): 28; and David Arnold, "Plant Capitalism and Company Science: The Indian Career of Nathanial Wallich," *Modern Asian Studies* 49, no. 5 (2008): 902.

5 William Griffith, *Posthumous Papers Bequeathed to the Honorable the East India Company, and Printed by Order of the Government of Bengal*, ed. John McClelland (Calcutta: Bishop's College Press, 1847–1848), ii; and William Griffith, *Report on the Tea Plant of Upper Assam* (Calcutta: n.p., 1840), 2.

6 Andrew B. Liu, *Tea War: A History of Capitalism in China and India* (New Haven: Yale University Press, 2020), 81; and Erika Rappaport, *A Thirst for Empire: How Tea Shaped the Modern World* (Princeton: Princeton University Press, 2017), 89–90. Today, Assam is India's largest tea-growing area.

7 Rappaport, *A Thirst for Empire*, 85; and Liu, *Tea War*, 32–33.

8 Griffith, *Posthumous Papers Bequeathed to the Honorable the East India Company*, ii. I am grateful to Tim Utteridge at the Kew Herbarium for bringing these tea specimens to my attention.

9 Lieutenant Colonel F. Wilford, "On the Ancient Geography of India," *Asiatic Researches; or, Transactions of the Society, Instituted in Bengal, for Enquiring into the History and Antiquities, the Arts, Sciences, and Literature, for Asia* 14 (1822): 424; and Barbara Maria Stafford, *Artful Science: Enlightenment, Entertainment, and the Eclipse of Visual Education* (Cambridge, Mass.: MIT Press, 1994), xxiv.

10 John McClelland and Thomas Horsfield, "A List of Mammalia and Birds Collected in Assam," *Annals and Magazine of Natural History* 6 (1841): 366.

11 John McClelland, "List of Mammalia and Birds Collected in Assam," *Proceedings of the Zoological Society of London* 7 (1839): 147 (emphasis by the author); and Griffith, *Posthumous Papers Bequeathed to the Honorable the East India Company*, ii. We do not know exactly where the hoolock in the Company painting was found during the tea expedition.

12 McClelland and Horsfield, "A List of Mammalia and Birds Collected in Assam," 366.

13 Griffith, *Posthumous Papers Bequeathed to the Honorable the East India Company*, ii; Lieutenant R. Wilcox, "Memoir of a Survey of Asam and the Neighboring Countries, Executed in 1825-6-7-8," *Asiatic Researches; or, Transactions of the Society, Instituted in Bengal, for Enquiring into the History and Antiquities, the Arts, Sciences, and Literature, for Asia* 17 (1832): 321, 323, 328–29; and Bikash Sarma, "'Tropicality' and Wildness: Experiential Travel Writing and 'Making Up' of Land and People in Nineteenth-Century Assam," *Asian Ethnicity* 21, no. 1 (2020): 59.

14 Frank Nicholls, *Assam Shikari: A Tea Planter's Story of Hunting and High Adventure in the Jungles of North East India* (Auckland: Tonson Publishing House, 1970), 84; and John Berger, *About Looking* (New York: Pantheon, 1980), 12.

15 Arupjyoti Saikia, "State, Peasants, and Land Reclamation: The Predicament of Forest Conservation in Assam, 1850s–1980s," *The Indian Economic and Social History Review* 45, no. 1 (2008): 81–87.

16 Maan Barua, "Plantationocene: A Vegetal Geography," *Annals of the American Association of Geographers* 113, no. 1 (2023): 14.

17 Maan Barua, *Plantation Worlds* (Durham: Duke University Press, 2024), 187.

18 Sophie Chao, "Plantation," *Environmental Humanities* 14, no. 2 (2022): 362.

19 Anna Lowenhaupt Tsing, Andrew S. Mathews, and Nils Bubandt, "Patchy Anthropocene: Landscape Structure, Multispecies History, and the Retooling of Anthropology; An Introduction to Supplement 20," *Current Anthropology* 60, supp. 20 (2019): S191.

20 Regarding nonhuman animals as agents on tea plantations, see Arnab Dey, *Tea Environments and Plantation Culture: Imperial Disarray in Eastern India* (Cambridge: Cambridge University Press, 2018); and Barua, *Plantation Worlds*, 1–20.

21 McClelland and Horsfield, "A List of Mammalia and Birds Collected in Assam," 366. The drawings were presented to the British Library in 1856 when the library was still a part of the British Museum.

22 William Tronzo, introduction to *The Fragment: An Incomplete History* (Los Angeles: Getty Publications, 2009), 4.

23 George M. Barker, *A Tea Planter's Life in Assam* (Calcutta: Thacker, Spink, 1884), vi.

24 Letter written by David Scott to James Kyd, Calcutta, 1820s, PrP 14-0001, Library, Art and Archives, Royal Botanic Gardens, Kew.

25 A. Burrell, "Indian Tea Cultivation: Its Origin, Progress, and Prospects," *Journal of the Society of Arts* 25 (February 9, 1877): 202–3; and "Letter from the Committee of Tea Culture to W. H. Macnaghten, Esq., Secretary to the Government of India, in the Revenue Department," *The Journal of the Asiatic Society of Bengal* 4 (January 1835): 43. For more about Nathanial Wallich and David Scott, see Martin Krieger, *Nathaniel Wallich: Global Botany in Nineteenth Century India* (London: Routledge, 2022), 213–31.

26 Dey, *Tea Environments and Plantation Culture*, 42. Wallich identified Gardner's plants as *Camellia drupifera.*
27 Pradip Baruah, "Wild Teas of Assam and North East India," *Journal of Tea Science Research* 7, no. 7 (2017): 34–39; and Arup Kumar Dutta, *Cha Garam! The Tea Story* (Guwahati: Paloma, 1992), 62–63.
28 Dey, *Tea Environments and Plantation Culture*, 42; and George D. Bearce Jr., "Lord William Bentinck: The Application of Liberalism to India," *The Journal of Modern History* 28, no. 3 (1956): 238.
29 Dey, *Tea Environments and Plantation Culture*, 43; William H. Ukers, *All About Tea* (New York: The Tea and Coffee Trade Journal Company, 1935), 1:139; and Arnab Dey, "Bugs in the Garden: Tea Plantations and Environmental Constraints in Eastern India (Assam), 1840–1910," *Environment and History* 21, no. 4 (2015): 538.
30 Jayeeta Sharma, "Making Garden, Erasing Jungle: The Tea Enterprise in Colonial Assam," in *The British Empire and the Natural World: Environmental Encounters in South Asia*, edited by Deepak Kumar, Vinita Damodaran, and Rohan D'Souza (Oxford: Oxford University Press, 2011), 122.
31 Jayeeta Sharma, "British Science, Chinese Skill and Assam Tea: Making Empire's Garden," *The Indian Economic and Social History Review* 43, no. 4 (2006): 437; John Lindley, "Substances Used as Food, Illustrated by the Great Exhibition," in *Lectures on the Progress of Arts and Science, Resulting from the Great Exhibition in London, Delivered Before the Society of Arts, Manufactures, and Commerce* (New York: A. S. Barnes, 1856), 174; and William Griffith, quoted in Ukers, *All About Tea*, 1:140.
32 Jayeeta Sharma, *Empire's Garden: Assam and the Making of India* (Durham: Duke University Press, 2011), 30–31. It is worth noting that doubts were also raised about Assam as a suitable place for growing "good" tea: "There was little probability of growing really good tea in such a jungly, warm region." John Lindley, "Professor Lindley on Substances as Food," in *Lectures on the Progress of Arts and Science, Resulting from the Great Exhibition in London, Delivered Before the Society of Arts, Manufactures, and Commerce* (New York: A. S. Barnes, 1856).
33 Sharma, *Empire's Garden*, 35; and Dey, *Tea Environments and Plantation Culture*, 45.
34 Sharma, *Empire's Garden*, 31, 35–40, 72–75; and Nitin Varma, *Coolies of Capitalism: Assam Tea and the Making of Coolie Labour* (Berlin: De Gruyter, 2016), 124–25.
35 Harlan, "Description of a Species of Orang," 53; James Rennie, *The Menageries: The Natural History of Monkeys, Opossums, and Lemurs* (London: Charles Knight, 1838), 1:174; L. C. Rookmaaker, "J. N. S. Allamand's Additions (1769–1781) to the *Nouvelle Edition* of Buffon's *Histoire naturelle* published in Holland," *Bijdragen tot de Dierkunde* 61, no. 3 (1992): 131–36; and Henry J. Noltie, "Zoffany's 'Monkey' and Other Early Depictions of the Western Hoolock Gibbon," *Archives of Natural History* 51, no. 1 (2024): 42.
36 Harlan, "Description of a Species of Orang," 52, 57.
37 M. L. Bose, *Development of Administration in Assam: With Special Reference to Land Revenue, Justice, and Police, 1874–1920* (New Delhi: Concept, 1985), 41; Amrita Shodhan, "The East India Company's Conquest of Assam, India, and 'Community' Justice: Panchayats/Mels in Translation," *Asian Journal of Law and Society* 2, no. 2 (2015): 360; Reeju Ray, "Interrupted Sovereignties in the North East Frontier of British India, 1787–1870," *Modern Asian Studies* 43, no. 2 (2019): 619–20; and Dilip Gogoi, *Making of India's Northeast: Geopolitics of Borderland and Transnational Interactions* (New York: Routledge, 2020), 58.
38 Burrough began his appointment in Calcutta in 1828. L. C. Rookmaaker, *The Rhinoceros of South Asia* (Leiden: Brill, 2024), 74; and "Marmaduke Burrough Papers 1829–1847," Mss.B.B946, American Philosophical Society, Philadelphia.
39 Achyut Kumar Borthakur, *The Administration of Justice in Assam (1826–1874)* (London: Routledge, 2019), 105–6.
40 Harlan's legacy would be embedded in the primate's identification as "Simia Hylobates hoolock. Harlan."
41 Harlan, "Description of a Species of Orang," 57, 59.
42 The fruit-holding hoolock is a female, whereas the other is a male that survived for a few months in Calcutta. The skull ("cranium") belonged to the male that perished at sea. Harlan, "Description of a Species of Orang," 57, 59; and B. C. Allen, *Assam District Gazetteers*, vol. 3, *Goalpara* (Calcutta: Baptist Mission Press, 1905), 29.
43 James Rennell, *The Journals of Major James Rennell, First Surveyor-General of India, Written for the Information of the Governors of Bengal During His Surveys of the Ganges and Brahmaputra Rivers 1764 to 1767*, ed. T. H. D. La Touche (Calcutta: The Asiatic Society, 1910), 57, 157.
44 Harlan, "Description of a Species of Orang," 57.
45 Yota Batsaki, "The Apocalyptic Herbarium: Mourning and Transformation in Anselm Kiefer's *Secret of the Ferns*," *Environmental Humanities* 13, no. 2 (2021): 394.
46 McCann, "Notes on the Colouration and Habits of the White-Browed Gibbon," 403.
47 Beth Fowkes Tobin, "Imperial Designs: Botanical Illustration and the British Botanic Empire," *Studies in Eighteenth Century Culture* 25 (1996): 275.
48 K. Sivaramakrishnan, "Colonialism and Forestry in India: Imagining the Past in Present Politics," *Comparative Studies in Society and History* 37, no. 1 (1995): 7.
49 Jardine's account borrowed heavily from Harlan and Burrough's descriptions of the primate. Jardine, *The Naturalist's Library*, 27:137–44.
50 The oak represented naval might, political power, wealth, property, and lineage. Martin Bell and Gordon Noble, "Prehistoric Woodland Ecology," in *Image, Memory and Monumentality: Archaeological Engagements with the Modern World*, ed. Andrew Meirion Jones, Joshua Pollard, Michael J. Allen, et al. (Oxford: Oxbow Books, 2012), 82, 85–86; Peter Young, *Oak* (London: Reaktion, 2013), 74–76; Sarah Couch, "The Practice of Avenue Planting in the Seventeenth and Eighteenth Centuries," *Garden History* 20, no. 2 (1992): 174; and Charles Mosley, *The Oak: Its Natural History, Antiquity and Folk-Lore* (London: Elliot Stock, 1910), 100.
51 Batsaki, "The Apocalyptic Herbarium," 395.

52 David Allen, "Tastes and Crazes," in *Cultures of Natural History*, ed. Nicholas Jardine, James A. Secord, and Emma C. Spary (Cambridge: Cambridge University Press, 1996), 400–403; William Wordsworth, "Nutting," *Pastoral Poems* (New York: D. Appleton, 1859), 30; Luke Keogh, "The Wardian Case: How a Simple Box Moved the Plant Kingdom," *Arnoldia* 74, no. 4 (2017): 6; and Griffith, *Posthumous Papers Bequeathed to the Honorable the East India Company*, 5–6.
53 Felix Driver, "Exploration as Knowledge Transfer: Exhibiting Hidden Histories," in *Mobilities of Knowledge*, edited by Heike Jöns, Peter Meusburger, and Michael Heffernon (New York: Springer, 2017), 90.
54 Nicholls, *Assam Shikari*, 83.
55 Swati Chattopadhyay, *Small Spaces: Recasting the Architecture of Empire* (London: Bloomsbury Visual Arts, 2023), 6.
56 Donna Haraway, *The Companion Species Manifesto: Dogs, People, and Significant Otherness* (Chicago: Prickly Paradigm Press, 2003), 5.
57 Pierre Bélanger, "Is Landscape Infrastructure?," in *Is Landscape . . . ? Essays on the Identity of Landscape*, ed. Gareth Doherty and Charles Waldheim (London: Routledge, 2015), 215.
58 Bélanger, "Is Landscape Infrastructure?," 216.
59 Nicholls, *Assam Shikari*, 84 (italics by the author).
60 Graham Livesey, "Assemblage," in *The Deleuze Dictionary Revised Edition*, ed. Adrian Parr (Edinburgh: Edinburgh University Press, 2010), 18 (italics by the author).
61 Harlan, "Description of a Species of Orang," 57–58.
62 Noltie, "Zoffany's 'Monkey' and Other Early Depictions of the Western Hoolock Gibbon," 37–46.
63 George Candler, "Notes on the Habits of the Hoolock," *Journal of the Bombay Natural History Society* 15, no. 4 (1904): 702; and Maan Barua, "Bio-Geo-Graphy: Landscape, Dwelling, and the Political Ecology of Human-Elephant Relations," *Environment and Planning D: Society and Space* 32, no. 5 (2014): 915–34.
64 Sujayita Bhattacharjee, "The Scenario of Man-Elephant Conflict in Hoollongapar Gibbon Wildlife Sanctuary of Assam, India," *International Journal of Scientific and Research Publications* 2, no. 8 (2012): 2; and Sandrine N. Dam, "A Short Study on Wild Hoolock Gibbons (*Hoolock hoolock*) in Assam and Bangladesh," *Gibbon Journal* 2 (2006): 41.
65 Thomas Geissmann, "The Gibbon Network," http://www.gibbons.de/main2/03fieldsites/india.html; Anwaruddin Choudhury, "The Distribution and Status of Hoolock Gibbon, *Hoolock hoolock*, in Manipur, Meghalaya, Mizoram, and Nagaland in Northeast India," *Primate Conservation* 20 (2006): 79–87; J. Das, J. Biswas, P. C. Bhattacharjee, and S. M. Mohnot, "First Distribution Records of the Hoolock Gibbon, *Hoolock hoolock Leuconedys* from India," *Zoos' Print Journal* 21, no. 7 (2006): 2316–20; Anwaruddin Choudhury, "The Distribution and Status of Hoolock Gibbon, *Hoolock hoolock*, in Karbi Anglong District, Assam, Northeast India," *Primate Conservation* 24, no. 1 (2009): 117–26; and Anwaruddin Choudhury, "The Hoolock Gibbon (*Hoolock hoolock*) in Tinsukia and Dibrugarh Districts of Assam, India," *Asian Primates Journal* 1, no. 2 (2009): 24–30.

6

BLOWOUT BEARDTONGUE

The Glitch in Landscape Conservation

ROSETTA S. ELKIN

> A glitch in any kind of system creates minor disturbances without actually damaging its major functioning. Glitches do not stop transmission; they merely make it scrappy, dirty, or noisy.[1]

Which plants matter to the making of human history? Value is commonly found in only the most productive or consumable part of each species—the pulp of palm fruit, ferments of agave, and bolls of cotton come to mind. These plants nourish cultural, economic, and political consensus as they make history. Other plants persist along the margins of human life, making history scrappy, dirty, or noisy, because they have no immediate worth or use. Thousands of plants are sidelined by history because they tend to disturb records, much like a *glitch*, a blip or bug that enters a system and shifts it, stimulating change. Such plants agitate more than they appease, and they do little to make history because they neither feed the planet nor solve our problems. I am referring to the spontaneous, unpredictable plants that underlie, orient, and constrain us.[2] The question of which plants matter to the making of human worlds is a significant one, because it suggests how historical interpretations impact populations well beyond the boundaries of any particular profession. Rather than resolving or deleting these *glitch* species from the record, I am suggesting that it is time to center living, behaving plants that have no material gain or immediate benefit to humankind. Even the funny-looking ones, the aggressive ones, the ugly ones, and the ones that stick out. Plants that slip through the cracks might encourage a more inclusive and thus equitable plant humanities.

This essay pays close attention to one particular glitch species that refuses to be boxed in by history. The rare perennial forb blowout beardtongue (*Penstemon haydenii*) is a federally endangered species that slips in and out of human worlds.[3] Ferdinand V. Hayden, mischievously known to the Sioux as "man who picks up stones running," was an enterprising geologist who first surveyed Nebraska and described the plant in 1857.[4] But this essay does not pay close attention to humans making history; rather, blowout beardtongue makes history, revealing that what is endangered is not a plant but a habitat misinterpreted and pacified by humans. I am referring to the mobility of the dune landscape and the urge to plant and fix it in place. I am also referring to the formation of blowouts that break the plant-anchored surface and expose persistent seed-banks. Blowout beardtongue loves a blowout. Or searches for a blowout, preferring the windward sandy slopes. The particular scrappiness of blowout beardtongue confounds conservation efforts because to protect the plant is to embrace a blowout, bewildering priorities. Since its entire global range is limited to the Nebraska Sand Hills, it denies the world of fixed settlement, possession, and resource partitions.[5] As its name suggests, the Sand Hills are both sandy and hilly, they are mobile and itinerant, and they help blowout beardtongue make history by virtue of the ceaseless human struggle to hunt, forage, pasture, hold, conserve, or otherwise settle amid the dunes.

To attend to blowout beardtongue on its own terms is to follow how conservation biologists grapple with a landscape contoured by a history of asymmetrical struggles between indigeneity and immigration, ranching and farming, drought and irrigation, spontaneity and management. Blowout beardtongue endures in its endemic range on its own terms, thriving outside of conservation mandates and despite restoration initiatives. Therefore, glitch is all about claiming the right to a history full of narrative complexity within traditions that tend to center humankind alone. Glitch is also about revealing how the inclusion of messy contexts, or landscapes, mediates a more equitable relationship between human and plant life.

It is too simple to state that blowout beardtongue is limited to the Sand Hills. It is more particular than that. Blowout beardtongue only emerges in a parabolic dune, a landscape formation described as a blowout. Parabolic dunes are secondary features that represent a younger history of the dune field, oriented in any direction because of the complex wind patterns.[6] Blowouts are deep depressions shaped drastically in time by strong winds that break the outward surface of grassy, plant-anchored dunes. The sheer force of a gust will crack open the shallows of a high dune, as sand is quickly exposed to a more and more erosive force that expands the ever-widening hollow. Blowouts help dunes shift and stay mobile. However, they disfigure attempts at settled agriculture and obstruct cycles of pasturing, since a blowout can be miles wide and up to twelve meters deep. Blowouts blur policy, softening hardened ownership lines. Blowouts take land offline, and only blowout beardtongue can take it back.

Herein lies the matter: blowout beardtongue loves a blowout. It is the only plant willing to enter into this risky relationship, and it does so on its own terms by resisting all human attempts to reclaim or "restore" a blowout bowl using cultivation techniques, transplanting systems, nursery standards, and even attempts to broadcast seed. All attempts fail. Planting blowout beardtongue would be very convenient for farmers and ranchers; it would lend some authority to conservationists aiming to keep up continuous grasslands. Instead, blowout beardtongue decides when and where it begins the slow process of spreading across the freshly exposed dune, sneaking around in unpredictable ways. Blowout beardtongue refuses to behave through our species' cultural and scientific inconsistencies, creating a glitch in the system—a two-foot-tall resistance that refuses to scale up, speed up, or grow up. Its behavior calls for a different kind of examination within the rules of conservation, one that decenters the human by acknowledging that many neglected or unstoried plants on the margins have, in fact, defined the center.

In 1982, Eric R. Wolf published *Europe and the People Without History*, a book that captured how scholars—historians, anthropologists, and sociologists—packaged societies and entire civilizations and continents into neat boxes for analytical purposes, portraying them as self-contained entities without connections to each other and to the larger worlds around them. Wolf's "people without history" was first and foremost a formulation of how European history is stunted and flattened by the failure to see the world as a whole, a totality, or an interconnected system: "If there are connections everywhere, why do we persist in turning dynamic, interconnected phenomena into static, disconnected things?"[7] Wolf contrasted the largely scholar-civilized West to an quintessential East, where life was cheap and education was meager. Inevitably, these two categories and the frictions between them yielded to the rise of "Third World" underdevelopment, a contrast to the developed West and the developing East. Such categories, Wolf asserted, are the building blocks of the social sciences, an entirely biased categorization that he referred to generously as limited, preferring to reference "bundles of relationships," a phrase that resonates with the unruly topic at hand.[8] Is land conservation stunted and flattened by the failure to see the landscape as a whole, a totality, or an interconnected system?

Thinking with Wolf captures how landscape architecture, conservation science, and restoration ecology often package other species into neat boxes, identifying a method for the plant humanities where plants and people are not split by boxed-up intellectual justifications but driven by the same empathy for context that promotes a more sensitive understanding of the world.[9] Using Wolf's logic, there can be no "plant history" apart from "human history," only a component of a collective memory that is suppressed or omitted from more established accounts. I am suggesting that what is left out are the manifold plant species that are not immediately useful to human worlds—those without medicinal value, exploitable fruit, resin, or seed. Such a framing also unearths the reified classification between "productive" and "nonproductive" plant species, those that might not retain an erosive slope or take well to containerized standards. Those plant species—like blowout beardtongue—that thrive on their own terms conceal themselves with increased human attention and wither through historical processes that help make human worlds.

The glitch reveals how blowout beardtongue prospers and why conservation fails across the Sand Hills. Advocating for a closer reading of plant life is a way into a more sensitive environmental ethic based on coexistence and adaptation across the landscape.[10] There is no center when learning from plants that are not economically important but tend to leak from one sector into another in their full aliveness. When plants are stripped of their place in culture, and their relations are reduced to standards, they lose their complexity. Planting as a type of conservation becomes a purely symbolic pattern when plants misbehave. Glitches can only appear in these contexts, reasserting complexity in the normative techniques of domination. Glitches crack open the code.

The first half of this essay considers blowout beardtongue through its growth and development, following naturalists in their attempt to discover how it spreads seed, reiterates belowground, collaborates with the microclimate of blowouts, and thrives on the strata of disturbance. My ambition is to center "how plants grow" through their evolutionary history rather than default to how they look or what bargain they might offer humans. Some attention is necessarily paid to the landscape through the spatial formation of blowouts and how they advance plant cycles temporally. The second half takes a walk through the Sand Hills to explore its scale, considering that they are the combined size of Massachusetts, Rhode Island, and Connecticut. Research shifts constantly from the intimacy of a seed or a grain of sand to the regional and macro scale of the Ogallala Aquifer and back again, allowing us to see how even the largest global processes of colonialism are embedded in local settings where they are laden with new meanings. A distinct and unusual landscape like the Sand Hills requires some decoding in the present since existing ecology is often the result of past human activities.

I use the word "landscape" in its fullest sense by emphasizing interconnections. Here again, I am inspired by Wolf's determination to decenter humans, suggesting that the bond between the physical qualities of the land and the behavior of plant life can also forge transcending networks of power. These are the bundles of relationships that cocreate what humans call landscape in the immutable space where the living plant meets the living soil at zero degrees and zero elevation. At ground level, monitoring is scrappy, calculation is dirty, and accuracy is noisy—and yet, the scale of relations underfoot is the most ordinary expression of the earth because it holds no visual, spatial, or social coding. The landscape is where blowout beardtongue lives and breathes, forcing upon human worlds more intimate ways of making history.

Blowout Beardtongue

Penstemon is a compound word from the Greek word *pente* (five) and *stemon*, a charming reference to a fifth stamen that protrudes from the flower and looks hairy enough to be called beardlike.[11] A stamen is the male part of the flower, but this fifth stamen is distinguished by being sterile or pollen-free compared to the four others. *Penstemon* is described ecologically as the largest genus of flowering plants endemic to North America; horticulturally, it is an "impossible" plant to breed and domesticate, and it is labeled short-lived but drought-tolerant by gardeners.[12] As Robert Nold remarks in his book on the genus, the fact that most Americans descend from an agricultural heritage means that we inherit a tendency to suffocate plants by "working organic matter into the soil, fertilizing and watering" rather than appreciating that many plants evolved with low nutrient levels and thrive with a greater degree of inattention.[13] Still, there are as many assumptions about this genus as there are specialized habitats across its range, from tiny cushion-like formations that hug alpine rocks to towering individuals that persist in prairies. Consider that *Penstemon haydenii*, or blowout beardtongue, is only one of roughly 280 species in the genus. Such morphological diversity deserves to be a little unruly, a little less ruled.

The scientific names of plants tie centuries together, but common names also appear through discovery and accumulation. Beardtongue is the common name for *Penstemon*, but there is no strict rule for what to call the plant, since the scientific and common names are used interchangeably. We could easily call this plant blowout penstemon and no one would flinch at the lack of italics. There are other aptly named *Penstemons*, including beardlip, sandloving, fuzzytongue, closethroat, and littlecup that all feel like a good starting point for everyday agreement. Liberty Hyde Bailey offers a good reminder of the attraction of common names: "The naming of plants under the rules of nomenclature is an effort to tell the truth. Its purpose is not to serve the convenience of those who sell plants or write label or edit books; it is not commercial. Serving the truth thereby serves everybody. In the end, nomenclature rests on the plants rather than on printed regulations."[14] Which is why I'll stick to beardtongue instead.

Blowout beardtongue is a herbaceous perennial that grows in large, multistemmed clumps. These clumps hold roots and shoot buds and leaves reaching one to two feet in height, with floppy stems that curve and touch the ground (or sand, as it were). The vegetative stems hold unusually narrow basal leaves and remarkably oversized floral bracts. One collector called its asymmetry "grotesque," a judgment on how top-heavy the plant appears above ground.[15] Belowground is where blowout beardtongue really lives, as stem nodes forage for resources and the plant buries itself deeper and deeper into the sand until root hairs find the slightest moisture and signal to tempt mycorrhizal relationships. Mycorrhizal fungi are microorganisms that live concealed in the substrate and support a number of plant functions, from water absorption to chemical communication.[16] Despite the expansive hidden geography and spread of blowout beardtongue, woody roots are not particularly good at actual absorption. To overcome this, each micro-movement of the root tip proliferates cells and reaches out to form relationships in the interest of absorbing moisture. The ability to extend beyond the "boundary" of the root in order to engage with absorption transpires through root hairs that attract bacteria, fungi, and other microbes into a mutually beneficial relationship with the rest of the plant. Particular to this relationship are the mycorrhizal fungi that greatly enlarge the available soil volume for absorption through extensively fibrous networks. Once the moisture is absorbed, the plant reiterates by sending down some anchor roots before it curves again or grows upward. Repeat.

The leaves are opposite, as botanists say, but they are so much more: they lean and droop and

Illustrations generated by Rosetta S. Elkin, contrasting blowout beardtongue (*Penstemon haydenii*) and various blowout conditions across the Nebraska Sand Hills. Images from Rosetta S. Elkin and the USDA archives; used with permission and appreciation from Richard Gilbert, Nursery Manager, Nebraska National Forests and Grasslands, and Greg D. Wright, Grassland Wildlife Biologist, U.S. Forest Service.

look relieved if such a thing is possible. These loose leaves are greenish-blue or bluish-green, depending on the time of day, and reflect the sky on their waxy, hairless surfaces. When a flowering stem emerges, it is straight compared to the vegetative stems, and rises to eighteen inches tall on a thick, juicy stalk as wide as half an inch in diameter. Its distortion is its drama, its beauty.

Blowout beardtongue is considered a colonizing species, which means that it reproduces asexually by means of vegetative offspring that remain attached to the parent, at least until they establish. That is how it once appeared to early conservationists working the landscape. At this time, blowout beardtongue was considered so volatile and rare that it was seldom dug up when it was found, and because it was never unearthed, the liveliness of its life *in* the Sand Hills was overlooked. The clonal hypothesis endured until a study exposed the risks of overcounting plants with clonal reproduction: "Carefully excavating sand around ten pairs of stems, we found that the subterranean axis was vertical in all cases and none of the plants were connected underground, even those that were only 15 cm apart."[17] What this means is that blowout beardtongue is that plants are germinating from seed, evolving to resist the lack of blowouts and waiting it out underground—another glitch in the system.

Billions of seeds are obscured in moving sand over hundreds of years, while others are scattered within a few days after dispersal. Seeds are the object of study now, revealing thick seedcoats that ensure long periods of dormancy in those that are deeply buried.[18] For the reader, consider replacing the image of large clumps actively moving the aboveground plant from below in a slow, concealed crawl within the blowout, with attention to ancient seeds waiting it out underground, and supporting the endurance of the plant across time and against the odds. According to nursery managers and botanists working in the Sand Hills: "Unlike most *Penstemon* species, blowout penstemon seeds have narrow, thin wings. Most seeds fall near parent plants, and remaining seeds are dispersed by wind, ants, small rodents, or birds. Seedcoats are scarified by blowing sands and germinate during late spring or early summer or when deeply buried seeds are brought to the upper few centimeters of sands."[19]

Seed dispersal is one of the least understood phases in plant life cycles, despite being recognized as fundamentally important for understanding plant distribution, abundance, population dynamics, and life-history variation. For species living in fragmented habitats, like the Sand Hills, seed dispersal is crucial for regional persistence. In the seed stage, blowout beardtongue burrows and forms significant seed banks at depths that protect it from ranchers, farmers, and conservationists, as well as from insect predation and overgrazing. Imagine multiyear stockpiles of seed resources awaiting favorable climate conditions, a cryptic stage in blowout beardtongue's life history that unfolds in concealed layers, far from humans until a blowout forms.

When seeds are ready, they are slowly distributed in tandem with the climate—as winds shake, its grotesque stalk moves like a baby's rattle, shedding and dispersing seeds directionally along varying patterns.[20] Once seeds land in the blowout, or on exposed sand, the next gust or storm moves them even further. But it does more: the sand scarifies or sandblasts their coating so that each seed is primed for growth. Then the plant dies back and the seeds live on. Blowout beardtongue is quite short-lived above ground, as a plant, but scientists believe that seed stocks can remain viable underground for decades. Imagine that when a blowout occurs at the landscape scale, it is not breaking the grassy surface, but revealing dormant seeds, exposing the next generation to light and moisture. In time, we can imagine that the proven terms of months, years, or decades in research reports might also just scratch the surface.

According to Sand Hills conservation director Greg Wright, a U.S. Forest Service biologist, blowouts used to be hundreds of meters wide, miles even, undulating between five and twelve meters in depth. Blowouts were a regular occurrence that helped move the sand sea along. But the reduced grazing intensity created by the implementation of grazing systems by ranchers, as well as wildfire suppression over hundreds of years, has resulted in increased plant cover that stabilizes blowouts. Blowouts represent poor ranching and ruthless farming; they are a cultural nonstarter. With the decreased frequency and extent of habitat for blowout beardtongue, or the success of conservation, so-called unstable sand dunes are under control. And with this control comes the label of extinction, projected on a plant that shakes it off every time it disperses more seed. Humans tend to use the word *hate* in relation to disturbances of the unplanned variety, like blowouts. But blowout beardtongue might just change our minds; it might just teach us to love blowouts again.

With so much seed at hand, or underfoot, it remains a mystery as to why this plant should be considered endangered, which is why all attempts to refine nursery standards are critical. Seed is scarified by hand and placed into well-ordered trays, hooked up to drip-line that feeds plastic pots filled with carefully diversified soil inside the greenhouse at Bessey Nursery.[21] Plugs are successfully grown and brought out into the field, as field crews get started planting blowouts created by conservationists. These designed blowouts are created to mimic habitats, as conservation teams wait for massive seed stores to emerge. According to Wright: "Our approach is scattered risk. Because we are inevitably wrong." Here, the glitch pulses with life, as hundreds of carefully raised plugs are planted out and pounds of seed are broadcast across these artificial blowouts. Years pass and nothing happens. Despite all the research and effort, the mysterious timing between blowout dynamics and blowout beardtongue that initiates succession and invites more plants, pollinators, and creatures into their ecology cannot be mimicked. Fast forward across thirty years of conservation efforts aimed at creating acceptably small blowouts for the beardtongue by manipulating cattle to congregate near certain water tanks where the plant is growing. And still nothing.

Currently, with the cultural shame and discord afforded to them, blowouts created by Wright and this team cannot exceed a few meters in width. To quote Wright: "What is truly missing is scale, for now all we can do is garden the blowouts." What he means is that the plants are not regenerating naturally, they rely on humans or they wither. The plant breaks apart the divisions between horticulture and ecology, between cultural and biological conditions, ranching and conservation, propagation and extinction. No blowout, no beardtongue.

Nebraska Sand Hills

The Nebraska Sand Hills provide the setting for a description of blowout beardtongue, a physical region that distinguishes itself from the rest of Nebraska and much of the Great Plains. They are the largest dune field in the Western Hemisphere, covering an area of more than 57,000 square kilometers (14 million acres).[22] Within the regulated agri-surface of the state, the Sand Hills ripple with dunes hemmed in by a range of bluffs, breaking the regularity of the gridded plots that surround them. Yet, the region endures almost unscathed from its formation in the late Pleistocene and remains full of these kinds of contrasts: physical cohesiveness and cultural complexity, unscathed prairie and artificial forests, expansive dryness and sizeable water resources. The plant associations echo its distinctive margins, as grass-packed sand dunes vary from green to brown to a glowing orange depending on periods of drought. Dunes are made choppy by sporadic wind erosion, relatively low precipitation, and temperature extremes where exposed elevations have thinner surface transitions and support drought-resistant plants such as prairie sandreed (*Calamovilfa longifolia*), sandhill muhly (*Muhlenbergia pungens*), and hairy grama (*Bouteloua hirsuta*). Prominent stands of soapweed (*Yucca glauca*) iterate sporadically, while the blowout beardtongue abbreviates the steep inclines limited to blowouts. The encounter with grassland dune plants in the Sand Hills is one of endless transitions where plant behavior is expressed topographically and temporally.

What is radical is that the Sand Hills will not be erased or rubbed smooth by a few hundred years of attempts to embed settlement. The mobile dunes are certainly one cause, and impossible to overlook. But it is the totality of the undulating landscape that expresses the substrate, since the dunes register the major recharge zone for the Ogallala Aquifer. The Sand Hills are simply the earthly expression of this ancient buried source. On the surface, the Sand Hills contour the movement of the aquifer, and plant life conveys the

connection between this deep moisture and the marshes, meadows, and fens that blanket the surface and migrate with the same pulses. The primary cover is grama grasses (*Bouteloua* sp.), which help stabilize areas that rise across the state of Nebraska in elevation of 1,880–3,600 ft in an east–west direction.[23] It feels like a sand sea, whereby the submerged, concealed thickness of the landscape provides ideal conditions for plant life because sand creates sharp drainage and freedom of movement for varied root systems. The landscape of the Sand Hills is not to be confused with desert dunes; rather, the topography undulates with verdant grasses and occasional clumps of trees, especially along the rivers that casually provide for remarkable diversity. I learned a lot about the Sand Hills from archival research, but nothing prepared me for the firsthand experience of moving through the landscape—of being in the Sand Hills.

From a faraway glance, the Sand Hills landscape does little to stand out visually from the rest of the Great Plains, expressing a homogenous and uniform pattern. The margins seem to blend from east to west, while interstate roads travel lengthwise in a north–south direction. Few noticeable boundaries arise until the eyes adjust to the monotony of aerial photographs or computer screens. With respect to those readers who are curious enough to find out more, the eastern margins are formed by a series of hills that rise above the rugged surface. Rivers are traceable as flats and fingers that interweave the margins, especially along the North Platte and Platte Rivers in Dawson County. The northern margin is marked by canyons, where sand dunes rise abruptly and expose rocky mounds that emerge from the Niobrara Valley, the so-called biological crossroads of the Great Plains. As Charles Barron McIntosh so aptly puts it: "human occupancy rather than sand sheets may determine the geographic boundaries along some margins of the Sandhills."[24] It is easy enough to survey this logic from afar, just follow the Niobrara River past the town of Valentine, as its pine-clad slopes give way to the agitation of crop circles.

In relation to human encounters in the Sand Hills, McIntosh proposes the concept of "sequential occupancy," as a way of recognizing the breadth of cultural practices across eleven thousand years: "Large numbers of people meant camps, camp residue, and cooking fires, and piles of bones often included lost projectile points. Bones and ashes that remained after animal-kill processing were frequently covered with alluvial or aeolian deposits and thereby preserved to the present time."[25] Sequential occupancy embraces an evolutionary perspective, but it does little to preserve the cultural traits and meanings of the plants that are equally successive. *Penstemons* are simply part of the North American landscape, outside of the artifacts used to build human chronology. *Penstemons* continue to provide medicinal use, just as they were relied upon for thousands of years by Native Americans. Extracts from the roots were applied to gunshot and arrow wounds, inserted into the tooth cavity for pain, and taken directly for consumption and whooping cough. The flowers were boiled to make a sweet tea, while the whole plant can be boiled into a poultice for headaches.[26] Because *Penstemons* live across the continent from Mexico to Canada, their lives are stored in thousands of recipes and stories that endure from protohistoric periods before the Europeans through the decades of Indian removals and the relentlessness of fixed settlements today.

Yet, to the registers of history, any consideration of the various tribes in the Sand Hills is necessarily an encounter that ends with the escalation of violence. These same registers cite trade contact close to 1700, when the Comanche dominated the western ranges and drove the Plains Apache-Padouca from the region.[27] Human-to-human struggles intensify in a narrative of military encounters, as the sandy margins of the Sand Hills were populated by non-native squatters, federal outposts, ranchers, and religious leaders and ruled by acts and policies that demonstrated the role of land management as a mediator of social relationships. In such a perspective, it is difficult to view any given account as bounded, which is why there can be no "human history" without "plant history." What will be required in the future is not only to deny that separation but to understand more precisely the value of plant knowledge in everyday life.

The Sand Hills evolved with plants, progressing the movements of antelope, deer, bison, elk, and buffalo as hills and valleys were browsed and traversed by well-defined routes that contoured the Sand Hills.[28] These trails existed well before the inventions of geometry, when journeys were limited and wayfinding arrangements could appear and disappear over several generations. Sand Hills living was largely seasonal, as relationships flourished for centuries through game hunting, with few settled practices or subsidence communities.[29] It was not until the years between 1870 and 1874 that surveys "discovered" the Sand Hills and named plants after prospectors, marking boundaries, rivers, and springs for private ownership. This is the culture of possession that some plants resist. In this short history, the most acute moments unfolded when ranching was incentivized. The expansion of horses and cattle necessarily includes the concomitant introduction of warm-season grasses, including big bluestem (*Andropogon gerardii*), sideoats grama (*Bouteloua curtipendula*), and indiangrass (*Sorghastrum nutans*). Ultimately, the Kinkaid Act (1904) facilitated access to the region, as land grants increased from 160 to 640 acres due to the lack of cover for feeding animals: "The grass is so scanty that the herdsman must have a large area for the support of his stock."[30] Along with these takings came the planting of trees for timber claims. Consequently, colonization was not only an act of erasure, it was an accumulation triumph, the enlargement of a stock of plants useful to settlement, and the complete disregard of those deemed useless or unproductive.

Here, I turn to a number of authors, including Winona LaDuke, Roxanne Dunbar-Ortiz, Dina Gilio-Whitaker, and Pekka Hämäläinen, who are reworking the terms of a strictly colonial history and the convenient narrative of erasure, by framing the endurance of Indigenous populations prior to 1492 as powerfully resistant and ecologically creative, especially in regard to the inclusion of lands, peoples of all cultures, and other organisms in everyday life. Consider this statement by Dina Gilio-Whitaker: "In Indigenous worldviews, there is no separation between people and land, between people and other life forms, or between people and their ancient ancestors whose bones are infused in the land they inhabit and whose spirits permeate place."[31] There is a direct relationship between the loss of land and the loss of plant knowledge that makes a difference to our understanding of the worlds we inhabit in the present and the ways of being responsible in the future.

Afterword

Conservation is a word used and abused in the Sand Hills. It was used to reassure settler colonialism; it was used to advocate for taming the sand to build fences and roads; and it was used to insist that planting trees was the solution to the failures of fixed settlement. In the Sand Hills, the struggle to hold and demarcate land continues in the friction between savvy ranchers, short-rotation farmers, and restoration-minded stewards. If the question is, why does plant conservation fail to reproduce blowout beardtongue, then the answer is also species-specific. The glitch has nothing to do with conservation as such but becomes part of the terrain for overturning its assumptions.

In the afterword to *Europe and the People Without History*, Wolf states: "As we unraveled the chains of abuses and effects at work in the lives of particular populations, we saw them extend beyond any one population to embrace the trajectories of others—all others."[32] These "others" expand in the twenty-first century to include the aggregate relations between plants and people, not as bordered species but as interconnected processes. These "others" resonate with Indigenous worldviews. The "others" are often the species that resist commodification. Advocating for a reinterpretation of the past not only informs the present but achieves a more inclusive and truthful historiography.

The scrappy transmissions of blowout beardtongue are an indicator of a different kind of hope because plants without history are neither an anomaly nor limited to the Sand Hills. They are part of an earthly story where paths of converging evolution overlap with the propagation of cultural life. They are all around us, all the time. The glitch is when we notice an exclusion. In turn, we know something more about our own species.

Notes

1 The term *glitch* is deliberately used for its association with technological progress; it captures a disruption that is the cause and effect of failure, often used in contemporary media and feminist studies. Sean Cubitt, "Glitch," *Cultural Politics* 13, no. 1 (2017): 19–33; and Legacy Russell, *Glitch Feminism: A Manifesto* (London: Verso, 2020).

2 These three words—underlie, orient, and constrain—are taken directly from the analysis of social orders offered by Eric R. Wolf, *Europe and the People Without History* (Berkeley: University of California Press, 1982), 76.

3 Petitioned as a Category 2 by the USDI FWS in 1980 and formally categorized as endangered in 1987. Environmental Conservation Online System (ECOS), U.S. Fish and Wildlife Service, https://ecos.fws.gov/ecp/species/6172.

4 Mike Foster, *Strange Genius: The Life of Ferdinand Vandeveer Hayden* (Niwot, Colo.: Roberts Rinehart, 1994), 60.

5 Since the "recovery plan," three known populations are also known to exist in the Ferris Dunes of south-central Wyoming. Bonnie Heidel, Samuel E. Cox, and Frank C. Blomquist, "Dune Habitat Trends of an Endangered Species, *Penstemon haydenii* Wats. (Blowout Penstemon), in Carbon County, Wyoming," *Western North American Naturalist* 78, no. 2 (2018): 119–25, https://doi.org/10.3398/064.078.0202.

6 Thomas S. Ahlbrandt, et al., *Geologic and Paleoecologic Studies of the Nebraska Sand Hills* (Washington, D.C.: United States Department of the Interior, Geological Survey, 1980), 5.

7 Wolf, *Europe and the People Without History.*

8 Thomas Hylland Eriksen, foreword to Wolf, *Europe and the People Without History*, ix–xviii.

9 I am grateful to Dr. Anna Tsing for a conversation in 2019 that prompted me to reread Eric Wolf and begin to integrate the idea of including plants in the discussion of expanded or revised histories.

10 For more on plants and landscape, see Rosetta S. Elkin, *Plant Life: The Entangled Politics of Afforestation* (Minneapolis: University of Minnesota Press, 2022).

11 "Penstemon," Flora of North America, http://floranorthamerica.org/Penstemon.

12 Glenn Viehmeyer, *Penstemon in Your Garden* (Lincoln: University of Nebraska, College of Agriculture, Agricultural Experiment Station, 1961).

13 Robert Nold, *Penstemons* (Portland, Ore.: Timber Press, 1999), 11.

14 Liberty Hyde Bailey, *How Plants Get Their Names* (New York: Dover, 1963), 39.

15 David M. Sutherland, "Historical Notes on Collections and Taxonomy of *Penstemon haydenii* S. Wats. (Blowout *Penstemon*), Nebraska's Only Endemic Plant Species," *Transactions of the Nebraska Academy of Sciences* 16 (1988): 191–94, https://digitalcommons.unomaha.edu/biofacpub/34.

16 Anthony Trewavas, "Self-Organization and Behaviour in Root Systems," in *Plant Behaviour and Intelligence* (Oxford: Oxford University Press, 2014), 122–36.

17 Heidel, Cox, and Blomquist, "Dune Habitat Trends of an Endangered Species."

18 Kassie L. Tilini, Susan E. Meyer, and Phil S. Allen, "Breaking Primary Seed Dormancy in Gibbens' Beardtongue (*Penstemon gibbensii*) and Blowout Penstemon (*Penstemon haydenii*)," *Native Plants* 17, no. 3 (2016): 256–65.

19 Richard Gilbert, Tara Luna, and Greg Wright, "Propagation Protocol for Blowout Penstemon (*Penstemon haydenii*), a Federal Endangered Species of the Nebraska Sandhills and Wyoming," *Native Plants* 21, no. 1 (2020): 67.

20 The baby's rattle comparison comes from a conversation with Greg Wright, the lead botanist working to conserve blowout beardtongue in the Sand Hills.

21 Details of propagation are taken from interviews with Bessey Nursery manager Richard Gilbert, a biological scientist with the U.S. Forest Service.

22 Gilbert, Luna, and Wright, "Propagation Protocol for Blowout Penstemon (*Penstemon haydenii*)," 65.

23 For the most complete physiological and ecological history of the Sand Hills that includes human contact, see Charles Barron McIntosh, *The Nebraska Sand Hills: The Human Landscape* (Lincoln: University of Nebraska Press, 1996).

24 McIntosh, *The Nebraska Sand Hills*, 5.

25 McIntosh, *The Nebraska Sand Hills*, 9.

26 See, for instance, the twenty-six species of *Penstemon* and their manifold uses, as cited in the comprehensive volume by Daniel E. Moerman, *Native American Medicinal Plants: An Ethnobotanical Dictionary* (Portland, Ore.: Timber Press, 2009).

27 David J. Wishart, *Great Plains Indians* (Lincoln: University of Nebraska Press, 2016).

28 McIntosh, *The Nebraska Sand Hills*, 29; and Paul A. Johnsgard, *This Fragile Land: A Natural History of the Nebraska Sandhills* (Lincoln: University of Nebraska Press, 1995).

29 "It would appear safe to say that no prehistoric group entered the Sand Hills to practice agriculture as a primary subsident strategy." Ann S. Bleed and Charles A. Flowerday, *An Atlas of the Sand Hills*, Resource Atlas 5B (Lincoln: University of Nebraska, Conservation and Survey Division, 1990).

30 Arthur R. Reynolds, "The Kinkaid Act and Its Effects on Western Nebraska," *Agricultural History* 23, no. 1 (1949): 20–29.

31 Dina Gilio-Whitaker, *As Long as Grass Grows: The Indigenous Fight for Environmental Justice, from Colonization to Standing Rock* (Boston: Beacon Press, 2019), 27.

32 Wolf, *Europe and the People Without History*, 386.

TAXONOMIES

7

CIPOLLINO MEETS CESALPINO

The Adventures of Plants Amid the Traditions of Botanical Classification

ANATOLE TCHIKINE[1]

If the history of botanical scholarship were a balance sheet of losses and gains, it would show that the cost of scientific precision was the gradual rejection of the traditional forms of knowledge of nature. The development of classificatory principles illustrates this trend. What it reveals is a progression from ancient, medieval, and early modern "natural" taxonomies, based on the organic affinities among plants that supposedly reflected the "order of nature," to "artificial" systems including the Linnaean method, predicated on more abstract parameters (such as the number and arrangement of stamens and pistils), to contemporary systematics built around phylogenetic relationships, which are established on the basis of DNA sequences. A consequence of these epistemological shifts is the growing detachment of taxonomic groupings and schemes from everyday experience. Even in the case of the Linnaean classification, a flowering plant such as castor bean (*Ricinus communis* L.) is supposed to belong to the same class, Monoecia, as coniferous pines and firs, whereas what is now considered a single genus, *Verbascum*, happens to be spread over several different classes.[2]

There is probably no cause for alarm, since, historically, plant literacy had little to do with systematic botany. What needs to be recognized, however, is that we routinely classify plants in ways that are independent of scientific parameters. When we walk into a grocery store, for example, we do not think of fresh produce in terms of clades, families, or taxa any more than the Linnaean kingdoms, classes, orders, genera, and species. What we shop for are vegetables and fruit, as well as greens and herbs. There can be further subdivisions within these loosely defined categories. For example, we can identify berries, root vegetables, and tropical fruit, or look for heirloom or organic produce. The fact that these classifications have more to do with marketing strategies than with scientific divisions within nature is of little concern to us. Why should kiwis, for example, be considered less "exotic" than litchis, and why are cranberries seasonal whereas blueberries are not? It might, therefore, come as a surprise that tomatoes and eggplants are, strictly speaking, berries, while strawberries, by contrast, are aggregate fruits. As for bananas (which are also berries), they are only found at every fresh fruit counter because of their extraordinary success as a global commodity.[3]

Such commonplace groupings, even if of dubious accuracy, are not without merit. In a way, they help us orient ourselves among the conflicting messages of commercial publicity, family traditions, cooking habits, and nutritional values. Nor are these classifications redundant, because, unlike scientific taxonomies, they enable us to correlate our dietary choices with market typologies. This instrumental approach is hardly new. Early modern botanical authors habitually distinguished between native and foreign, edible and inedible, and curative and poisonous plants, grouping them in accordance with multiple, often overlapping and conflicting criteria. While such categories are now mainly the province of ethnobotany, it is hard to deny that they have greater diffusion than the more accurate, if obscure, scientific taxa. Recognizing this fact is not equivalent to admitting that science, by reaching further levels of complexity, makes itself irrelevant to the realities of everyday life. Rather, what such rival categorizations suggest is that scientific methods and data constantly have to compete with a wide spectrum of pre- and proto-scientific notions and attitudes that continue to shape our understanding of the natural world.

FIGURE 7.1
Raul Verdini, cover for the second edition of Gianni Rodari, *Le avventure di Cipollino*, 1957.

The premise of this chapter is that even though the loss of traditional forms of botanical knowledge was a precondition for the advancement of modern science, the result was their displacement rather than disappearance. In other words, rather than becoming erased, these epistemic systems, in Jürgen Habermas's terminology, retreated from the public to the private sphere, where, unregulated by scientific norms, they continued to be transmitted from one generation to the next.[4] While folklore, such as proverbs and sayings, culinary traditions, and alternative medicine arguably remain the principal repositories of this knowledge, another, less obvious source is children's literature.[5] Tapping into these resources affords us deeper insights into the cultural meanings of plants that are often intertwined with their taxonomic histories.

What follows is an attempt to explain the survival of such traditional—or, as folklorists call it, "vernacular"—botanical knowledge in the popular culture of twentieth-century Italy through the example of the afterlife of the medieval hierarchy of plants in Gianni Rodari's *Le avventure di Cipollino* (The Adventures of Cipollino) (1951) (Figure 7.1).[6] This classic of Italian juvenile fiction—a tale of oppression and resistance, betrayal and friendship, servitude and defiance—is the story of an onion boy in a world of anthropomorphized vegetables and fruit, who, driven by loyalty and courage, embarks on a quest for social justice as a champion of the exploited and the oppressed. Narrated with empathy and humor, Rodari's novel remains largely unknown to the Anglophone audience on account of its author's Communist affiliations.[7] These political sympathies are evident in the book's plot, in which the class struggle between the poor and rich results in a revolutionary upheaval that redefines the social order.

The protagonist, Cipollino, who leads the revolt, witnesses his father, old Cipollone, thrown into prison by order of the cruel governor, Prince Limone, the commander of an army of lemons of various sizes and ranks. Thus begins Cipollino's journey, which takes him away from his underprivileged urban household to a small, rural community nestled next to the castle of the affluent Countesses del Ciliegio, two conceited cherry sisters who bully their orphaned nephew, the trusty and studious Ciliegino. Cipollino helps a hapless old villager, Sor Zucchina (squash), and becomes apprenticed to the local cobbler, Mastro Uvetta (raisin). He also makes friends with other humble and industrious countryfolk, assuming the role of the defender of the poor and weak. His new associates include a musician, Pero Pera (pear), who plays a violin made from halved pear fruit; the good-natured market gardener Pirro Porro (leek), with the magnificent mustache that his wife uses to hang laundry; Zucchina's niece, the clearheaded and outspoken Sora Zucca (pumpkin); and the reclusive and empathetic Sor Mirtillo (bilberry), who is too timid to leave his tiny hideout in the woods.

Cipollino's main antagonist is the irascible and ruthless Cavalier Pomodoro (tomato), the majordomo to the countesses and administrator of their estate. Upon his first appearance, arriving in a carriage driven by four cucumbers, he is described as "a character of importance dressed in green, with a face so round and red that it seemed ready to burst like an overripe tomato."[8] In the ensuing confrontation, full of unexpected reversals of the protagonists' fortunes, Cipollino triumphs over his implacable foe with the help of loyal Ciliegino and their friends. Among them are peasant girl Ravanella (radish); Fragoletta (strawberry), a resourceful maid to the junior countess; young Fagiolino (common bean), son of the hustled ragman Fagiolone; and a pair of urchins, Tomatino (tomato) and Patatina (potato). They are joined by the testy but devout old mole Signora Talpa and the conscientious lame cricket Ragno Zoppo, a postman in the governor's prison. Their opponents and occasional, if unwilling, allies are Ciliegino's mean and pedantic tutor Don Prezzemolo (parsley); the cowardly village attorney Sor Pisello (pea); and two relatives of the countesses' late husbands, the gluttonous Baron Melarancia (orange) and the capricious and devious Dukeling (*duchino*) Mandarino (mandarin).

Cipollino's world is a parody of early twentieth-century Italy, a country in which rigid social hierarchies and fossilized feudal norms appeared side by side with militaristic ambitions and technological innovations. In Rodari's novel, such modern intrusions take the form of cinema, radio, telegraph, and electric lighting, as well as train services, public parks, and zoos. Traditional ratafia-flavored caramels coexist with imported American chewing gum, while the fear of microbial contagion through drinking water, expressed by guard dog Mastino, is as much at home as the predictably herbal and, therefore, Galenic cures administered by Doctors Fungosecco (dried mushroom), Nespolino (loquat), Carciofo (artichoke), and Professor delle Lattughe (lettuce).[9] A typical product of this transitional culture is Mister Carotino (carrot), a private detective hired by Pomodoro to search for the escaped prisoners. Armed with a vast arsenal of optical and photographic equipment, he uses it in such a disorienting manner that he is unable to leave his bedroom.[10]

These flashes of modernization, however, only offset the reactionary nature of Prince Limone's regime. The professional credentials of the medical experts attending to Ciliegino, for example, are judged primarily by their social standing. "A doctor without a carriage is a doctor without medicine," declare the countesses' servants when they try to turn Doctor Marrone (chestnut), a physician to the poor, away from the castle.[11] An improvised tribunal consisting of Pomodoro, Pisello, and Prezzemolo, while following mock legal protocols, imposes on the anguished villagers a tax on weather—standard on rain, double on thunderstorms, triple on snow, quadruple on hailstones, and a reduced rate on dew, fog, and frost—in addition to the existing duty on air that has already been declared the property of the countesses.[12] The culmination of these absurdities is the ghastly fireworks display that the prince puts on to entertain the Del Ciliegio sisters while trying to recapture their castle from Cipollino and his companions. In compliance with the governor's whim, rank-and-file soldiers are fired from cannons instead of pyrotechnics, decimating the battered lemon force.[13]

As this cast of characters suggests, the world that Cipollino rebels against is one where vegetables are dominated by fruit. This unequal status is evidenced by the dispute over a tiny cabin that a lowly squash, Zucchina, built on the land owned by the aristocratic cherries. The clear exception is violinist Pero Pera, whose profession—and, consequently, inferior social standing, judging by Mastino's supercilious reference to "low-class folk such as cobblers, orchestra musicians, onions, and the like"—was, however, a pun on the shape of pear fruit, a literary device analogous to Pomodoro's apoplectic complexion or Pirro Porro's spectacular mustache.[14] Yet why should Cipollino and his family—all "decent folk," as the author assures us in the beginning—end up at the bottom of the social hierarchy?[15] Why do lemons and other citruses form the core of the oppressive regime? Why is Pomodoro, despite his ostensibly noble rank and haughty attitude, in a subordinate position as the administrator to the countesses? Why is the inapt Sherlock, Carotino, unmistakably foreign, given both his title—Mister—and outlandish behavior? Why does the chestnut, Marrone (the physician who does not own a carriage), primarily treat the poor, unlike his wealthier colleagues: the dried mushroom, the loquat, the artichoke, and the lettuce?

As we try to answer these questions, we need to address whether any of these relationships can be understood as a form of botanical classification. What we seem to be dealing with is a hierarchy of plants that Rodari translates into a class structure or, looking at it another way, a social ladder projected onto the order of nature. The underlying hierarchical principle points to the premodern origins of this scheme. Is it accurate, however, to describe such hierarchy as an actual taxonomy rather than as a product of analogical reasoning of the kind that sixteenth-century Italian pundit Giovanni Battista della Porta expounded in his physiognomic theories?[16] This chapter attempts to explain the logic behind Rodari's choice of his fictional characters by analyzing the cultural trajectories of the corresponding plants through the lenses of taxonomic scholarship and dietary traditions of early modern Italy.

In a seminal article published in 1991, Allen J. Grieco maintained that medieval authors, in trying to bring order and meaning into the vegetal world, pictured it by analogy with the feudal social pyramid.[17] This model was based on the notion of the Great Chain of Being (*scala naturae*), which derived in part from the teachings of Aristotle. According to this cosmological doctrine, the universe was stratified into discrete tiers, with God and the angels at the top, humans in the middle, and the animal, plant, and mineral kingdoms as the three lower registers. The "chain" itself was the downward progression from the higher spiritual entities to other sentient creatures, endowed with a soul as the cause of motion and change, to the inanimate matter. Key to this system was the opposition of high and low, whereby trees, by virtue of being closer to the divine realm, were deemed superior to herbaceous plants, such as vegetables, grasses, and ferns. There is no reason to see this hierarchy as anything other than a form of taxonomic arrangement. In fact, one of the earliest botanical classifications that was put forward by Aristotle's student Theophrastus similarly divided plants into trees, shrubs, undershrubs, and herbs.[18]

It should be made clear that, from Theophrastus's perspective, these four classes had less to do with plant size and more with their growth habit and physical structure. This morphological approach gained further traction in early modern taxonomies. In the case of Italy—which remained at the forefront of botanical studies until the center of scholarly gravity moved north of the Alps, with the work of John Ray in Britain and Joseph Pitton de Tournefort in France in the late seventeenth century—a pivotal figure was Andrea Cesalpino, the author of the treatise *De plantis*

(On Plants) (1583).[19] Having received little attention in his lifetime, this book was declared a forerunner of the Linnaean method by none other than Carl Linnaeus himself because of the central importance that it attached to the reproductive organs.[20] While preserving the Theophrastian division of the vegetal world into four broad classes, Cesalpino's treatise was an attempt to put the inquiry into plants on a scientific footing by postulating a consistent set of criteria as the basis for taxonomic judgment.

The challenge for Cesalpino was the need to distinguish between the essential and accidental characteristics of plants. His philosophy of nature was firmly Aristotelian, in that it was based on a direct correlation between form and function or vegetal morphology and physiology. A plant, according to Cesalpino, was first and foremost a functional organism endowed with specific organs that enabled the physiological mechanisms of nutrition and reproduction. The problem with morphological features was that, when considered in their totality, they often pointed in contrary directions. The situation was further complicated by the presence of cultivars and the differences among them due to the divergent methods and conditions of cultivation. While Cesalpino admitted that some of the traditional groupings, such as cereals and legumes, were not strictly morphological, he systematically ruled out other criteria for species identification, including culinary and medicinal uses, location and habitat, and even taste and smell, as unrelated to the essential vegetal functions.[21]

What was left were the physical characteristics of plants—specifically, the parts considered responsible for nutrition and reproduction. Roots, which showed little typological variation, could not offer much by way of differentiating between the species. They were thought to be the only means of receiving nourishment until the discovery of photosynthesis in the late eighteenth century and the concurrent interest in carnivorous plants, such as Venus flytrap (*Dionaea muscipula* Ellis).[22] By contrast, the vegetal reproductive organs remained a source of constant fascination for Cesalpino. As he explained in one of his earlier writings:

> But admirable artifices and varied devices appear in the parts that enable generation, where it seems that it pleases nature above all to demonstrate to the world its beautiful works clad in various leaves, adorned with lovely flowers, and, lastly, carrying the most beautiful fruit, with such variety that seems endless among the multitude of species. This is why some plants display their seed almost naked, as do common fennel, giant fennel, and the like; others have it enclosed in various sorts of casings and vessels, some more simple, others more complex, some as single, others as multiplied in various modes and forms. Yet others, not so perfect, do not seem to produce any seeds, nor flowers or stems, as in the case of maidenhair fern, ferns, and the like; all of them have in place of seed fine fuzz covering the lower side of the leaf, where the plant's power to regenerate is contained, as has been observed in ferns that grow in places where they shed their leaves.[23]

This aesthetics of fructification, according to Cesalpino, was more than a display of nature's creative powers. It was a key to the identity of each species, as encoded in the number, position, and shape of its seeds and the structure of its seedpods. Built around these formal parameters, his taxonomic groupings set a new standard of botanical research, in which methodological rigor took precedence over the encyclopedic hunger that drove early modern studies of nature.

While aiming to produce a "natural" taxonomy, Cesalpino clearly gravitated toward a more "artificial" system. His push for coherent criteria, however, forced him to exclude such historically important aspects of plants as their places of origin, uses, and effects on the human body. This reductive method would find itself more at home with the Enlightenment community of experts than with the divergent chorus of Renaissance naturalists, who sought to reconcile Renaissance humanist scholarship and medieval herbalist practice to make sense of the growing body of empirical evidence.[24] It is hardly surprising, therefore, that, rather than creating a paradigm shift, Cesalpino's approach won little favor with his contemporaries, reluctant to part with their established modes of thinking. Relegated to relative obscurity, his book gathered dust until its rediscovery by Linnaeus in the mid-eighteenth century.

A more typical product of Cesalpino's professional and intellectual milieu was the monumental *Historia generalis plantarum* (General History of Plants), which was almost contemporaneous with his treatise. This compendium of botanical knowledge, published anonymously in Latin in 1586 and subsequently in French in 1605 under the name of the sixteenth-century Lyonese physician Jacques Daléchamps, is also known as the "Lyon herbal."[25] Unusually for such encyclopedic works, the book's organization presented a taxonomic arrangement, since each of the eighteen chapters was devoted to a particular grouping of plants. The sequence of these rubrics is no less astounding than the divisions within the vegetal kingdom that they marked, as indicated in the table of contents:

1. Trees that spontaneously grow in forests.
2. Shrubs that spontaneously grow among thickets and shrubbery.
3. Trees that are cultivated in gardens and orchards.
4. Cereal crops and legumes that grow in fields.
5. Vegetables and herbs that grow in kitchen gardens.
6. Umbelliferous plants.
7. Plants with pleasing flowers.
8. Odoriferous plants.
9. Plants that come from marshes.
10. Plants that grow in rough, stony, sandy, and sunny places.
11. Plants that come from shady, wet, humid, and fertile places.
12. Plants that grow in or near the sea.
13. Plants that climb on other plants.
14. Thistles and other thorny and prickly plants.
15. Plants with bulbous, fleshy, and knotty roots.
16. Purgative plants.
17. Poisonous plants.
18. Foreign plants.[26]

What was the intellectual basis for this bizarre taxonomy? An obvious parallel is the mock zoological groupings that Jorge Luis Borges, an Argentine writer of immense erudition, disguised as deriving from a pretend Chinese encyclopedia. Readers may be familiar with this fictional catalogue from his essay "The Analytical Language of John Wilkins" (1942) because of the uncontrollable laughter that it reportedly caused Michel Foucault, as evoked in the opening pages of *The Order of Things*. Borges's conceit, in the opinion of the postmodern theorist, was indeed a product of a different geography and age, since, characterized by sheer "impossibility," it possessed "the exotic charm of another system of thought."[27] This absurdist quality, according to the French thinker, was incompatible with the conventions of European scientific discourse, as codified by the succession of the three *epistemes*—the Renaissance, the Classical, and the Modern—that he postulated.[28]

What made the "Chinese encyclopedia" "impossible" for Foucault was the fact that Borges's paradoxical groupings did not share the same ontological plane and, therefore, could only belong to "an unthinkable space" (*heterotopia*).[29] In a similar way, plants in the Lyon herbal were classified in accordance with heterogeneous and overlapping parameters. These factors, in addition to morphological characteristics, included habitat, medicinal and dietary applications, and plant behavior, as well as olfactory and aesthetic factors. Such intellectual laxity, typical of the exponents of sixteenth-century natural history, was precisely the subject of Cesalpino's critique as well as a reason for his failure to garner sufficient support for his own taxonomy. Contrary to Foucault's conception of the successive *epistemes*, however, both classifications were products of the same intellectual culture, built on an uneasy and changing dynamic between the humanist and empiricist epistemological paradigms.

To conclude with Foucault that any classificatory system is ultimately arbitrary is not to say much. It is more important to emphasize that the advancement of botanical scholarship was not linear and that its historical trajectory depended on a complex interplay of multiple sources of knowledge. In the case of the Lyon herbal, some of its categories, such as "plants that climb on other plants," seem to be derived from popular reference works associated with sixteenth-century Italian encyclopedists Alessandro Citolini and Tommaso Garzoni, both operating within the intellectual tradition of the Renaissance "memory theater."[30] Their pedantic enumeration of various species in ways that they can be encountered and described was a lexicographic exercise rather than a means of streamlining the knowledge of nature. Such listings, however, were also an attempt to give structure to living experience, in which firsthand observation was refracted through the "period eye" of intellectual habits, professional practices, and daily routine.[31]

In light of this discussion, it becomes significant that neither Cesalpino's reductive taxonomy nor the looser groupings of the Lyon herbal draw upon the hierarchy of plants—an intellectual by-product of the Great Chain of Being, predicated on the cosmological opposition of high and low. Traces of this medieval doctrine abound instead in sixteenth-century horticultural writings. One example, notable for its empirical rather than speculative bent, is the unpublished *Agricoltura sperimentale* (Agriculture Based on Experience) (ca. 1593–1595) by Florentine Dominican friar Agostino del Riccio. This valuable manuscript contains frequent references to "noble plants" (*piante nobili*)—primarily, fruit trees—that are contrasted with "simples" (*piante semplici*) or medicinal herbs.[32] It would be wrong, however, to see this implicit dichotomy as equivalent to the social division between the nobility and ordinary or "simple" folk. Del Riccio's language, full of rhetorical inflections, can be notoriously flowery and imprecise, so that he can also describe an herbaceous plant, tobacco, as "noble" on account of its prestige as a New World novelty in addition to its allegedly curative properties.[33] As to the word "simples," its origin, rather than alluding to a humble or inferior status, lay in the ancient distinction between simple—that is, based on a single herbal ingredient—and compound medicinal preparations as used in Galenic medicine.

This evidence suggests that even though the medieval hierarchy of plants retained some currency in the sixteenth century, it was increasingly outmoded and pushed to the margins of what Foucault would regard as the mainstream scientific discourse. One reason for the slow demise of this scheme was that it had no bearing on the relative value of different species in terms of their utility and rarity, two fundamental criteria of early modern botanical judgment. An attractive Mesoamerican creeper with star-shaped scarlet flowers, now known as cypress vine (*Ipomoea quamoclit* L.), captivated the imagination of sixteenth-century students of nature more than the pedigree apples, pears, plums, peaches, cherries, and apricots long cultivated in Italian gardens.[34] The result was the gradual erosion of the notional boundary between "noble" and "simple" plants, no longer defined along the same lines as the Theophrastian distinction between trees and herbs.

What remained peculiar to orchard trees, however, was the social meaning of their fruit. Such produce, as Grieco has emphasized, was traditionally associated with an aristocratic diet and, conversely, was considered antithetical to peasant consumption. The reasons for this prescriptive attitude were physiological rather than economic. They lay in the ancient humoral theory, which constituted the framework for early modern medical knowledge. According to this doctrine, an illness was an imbalance of the four cardinal humors (blood, phlegm, yellow bile, and black bile), each characterized by a unique pairing of contrasting properties: heat and coldness, dryness and moisture. The excess of any of these fluids could be rectified by ingesting appropriate foods. Spices, which were considered dry and hot, for instance, could mitigate the effects of wet and cold phlegm, as manifest in the proliferation of mucus. In a similar way, some potentially unhealthy foods could be "corrected" with condiments and seasonings, being served in combination with other products. A classic example of this strategy was the "prosciutto e melone" (ham and melon) dish, in which cured meat supposedly neutralized the cold and wet gourd, thwarting its alleged propensity to "putrefy" in the stomach.[35]

The prevalence of one of the humors resulted in four basic personalities or temperaments (sanguine, phlegmatic, choleric, and melancholic), each marked by a predisposition for certain foods.

Added to this typology were occupation and lifestyle, since peasants toiling in the fields required different nutritional sustenance than aristocrats enjoying the pleasures of the hunt. Such contrasting diets were not simply a matter of choice, because human bodies were thought to partake in the properties of the food that they consumed. As a result, while elites found it hard to digest rough legumes (a notorious cause of flatulence), the more refined products served at courtly banquets were considered equally harmful to commoners engaged in regular physical work. In some apocryphal situations savored in sixteenth-century literature, such class-specific foods could even pose a mortal danger.[36] Eating habits, in other words, were fundamental to defining and sustaining early modern social identities.

An illustration of this stratified food culture is Annibale Carracci's late sixteenth-century painting of the "bean eater" (*mangiafagioli*) (Figure 7.2). It shows a farm laborer devouring a repast of stewed black-eyed beans, vegetable pie, bread, and green onions, accompanied by red wine, while he suspiciously eyes the intruding viewer. Carracci's canvas is typically construed as a social caricature, given the sitter's shabby clothing and ungainly manner, as he opens his mouth to slurp the broth dribbling from the raised spoon. What the critics fail to appreciate, however, are the dietary overtones of this scene. We are offered a formula for a healthy meal—the consumption of rough, hearty food as appropriate for robust stomachs. Accordingly, the man's coarse features, unruly hair, and bare chest glimpsed through the low-cut neck of his shirt convey the message of physical strength, as does the hand with unclean nails that rests on a piece of bread.

Having a bunch of green onions as part of this mise-en-scène was not a pictorial accident. Since classical antiquity, indulging in such foods, especially if eaten raw, was a marker of non-elite status, as was the "garlic" and "onion" breath that they left. Roman playwrights and satirists, from Plautus to Persius and Martial, ridiculed this habit as a sign of abject poverty or lowly standing, imbuing it with social meaning.[37] In a similar vein, although writing from a perspective of a different age, sixteenth-century Florentine artist and poet Agnolo Bronzino commented on the pervasive aroma of cooked onions as the smell distinctive of poor neighborhoods, as in the case of San Gallo and San Frediano in Florence, two urban areas associated with working-class households.[38]

Bronzino's burlesque eulogy of the onion captures the inherent ambiguity of this plant as a versatile and desired flavoring, indispensable in both cooking and medicine, but also as an attribute of lower social class. Such enduring attitudes explain Rodari's decision to make Cipollino the main character of his politically charged novel. They also account for the occupation of the leek Pirro Porro, a fellow plant of the genus *Allium*, whose job as a market gardener was probably analogous to that of Carracci's "bean eater." The character of Doctor Marrone, also associated with the poor, was informed by the same logic. Although a cultivated species, chestnuts were forest rather than orchard trees that usually grew in secluded mountainous areas, where they were harvested by local communities. Now primarily a winter treat that is served roasted by Italian street vendors, their fruit was an essential source of nutrition for the rural poor, especially in times of famine.[39]

The most complicated among Rodari's personages in terms of his social identity is arguably Cavalier Pomodoro. A North American plant treated with suspicion for much of the early modern period, the tomato's meteoric rise to the center stage of Italian cuisine during the nineteenth century was nothing short of remarkable. This success was due to a combination of several factors, among which changing medical theory, agronomic advancements, culinary innovations, and the rise of national consciousness each played a contributing role.[40] With his "ennoblement" being a recent phenomenon, however, Pomodoro was not on par with the hereditary aristocracy, such as his employers, the cherries. This newcomer status and the resulting social insecurities explain his resentment toward the parasitic castle guests Baron Melarancia and Dukeling Mandarino—two good-for-nothings, whose antics he despised but was expected to tolerate as the countesses' right-hand man.[41]

Another plant of Mesoamerican descent that owed its fortunes to nineteenth-century agronomic practices was zucchini or courgette, a green

FIGURE 7.2
Annibale Carracci, *The Bean Eater*, ca. 1584–1585. Galleria Colonna, Rome. Photograph © Stefano Baldini / Bridgeman Images.

squash of the genus *Cucurbita*, distinguished by its dark color and elongated shape. Recent research traces the beginning of its commercialization to Milan, from where it spread around the 1850s.[42] Rodari probably chose this cultivar, ubiquitous in his day as a market and garden product, to contrast Sor Zucchina's long and lanky physique with the plump, rounded forms of his niece Sora Zucca, the pumpkin—so stout that, during a prison escape with Cipollino and his friends, she had to be rolled downhill.[43] The comic effect of this pair, analogous to the grotesque duo of Melarancia and Mandarino, reflected the variety of shapes and forms in which squashes came. Although some of the first American plants naturalized in the food culture of Italy, they were considered under the same rubric as bottle or calabash gourds, long grown in Europe, and, accordingly, were broadly divided into two types: round and long.[44]

On the other end of the social spectrum and having their origins in the Old World, oranges and lemons, unlike their close relatives citrons, were also surprisingly recent additions to the horticulture of Italy. The arrival of these plants through the mechanisms of Middle Eastern trade had a transformative effect on late fifteenth- and early sixteenth-century Italian gardens, elevating citruses to almost an unrivaled status among other orchard trees. Sicily, with its strategic geographic and commercial position in the Mediterranean, was the first to succumb to this frenzy. Writing in 1510, Sicilian author Antonino Venuto extolled the orange as "clearly the king, prince, and lord of all trees, among which it is the most noble and shapely."[45] This sentiment was shared by Neapolitan humanist Giovanni Pontano, who proclaimed the plant's fruit none other than the golden apples of the mythological Hesperides.[46] While the reign of the orange may have come to an end by the turn of the seventeenth century, the passion for citruses was yet to reach its peak. A testimony to this enduring interest and to the longevity of

FIGURE 7.3
Ligurian or San Remo lemon, from Giovanni Battista Ferrari, *Hesperides sive de malorum aureorum cultura et usu* (Rome: Hermann Scheus, 1646). Photograph courtesy of the Research Library, Getty Research Institute, Los Angeles.

Pontano's metaphor was the treatise *Hesperides sive de malorum aureorum cultura et usu* (1646) by Jesuit scholar Giovanni Battista Ferrari, with separate chapters dedicated to citrons, lemons, and oranges. Illustrated with eighty botanical plates in addition to several allegorical scenes, it methodically listed different varieties of citrus trees to explain their medicinal and culinary uses and cultivation techniques, such as watering, grafting, and pruning (Figure 7.3).[47] Rodari's army of lemons echoed the privileged place of this plant in the garden and market cultures of Italy, fueled by the lasting popularity of its valued and admired fruit.[48]

This parade of botanical imports only offsets the puzzling foreignness of Mister Carotino, as attested to by his title of "Foreign Military Counselor" granted by Prince Limone during the siege of the Del Ciliegio castle.[49] Wild carrots were native to Italy, where they were cultivated and used mostly as a winter salad vegetable, usually cooked before being served.[50] No early modern Italian recipes, however, seem to treat them as an essential ingredient; nor are they listed among the standard produce—lettuce, squashes, cucumbers, common beans, chickpeas, radishes, parsnips, fennel, watermelons, onions, and melons—grown by the greengrocers of Prato, a Tuscan city known for its successful market gardening business.[51] Carrots acquired popularity in seventeenth-century Netherlands, where they allegedly even changed color from the more traditional yellow, white, or purple to orange in honor of the reigning House of Orange, which also ruled Britain between 1689 and 1702. Subsequently, having left their mark on Victorian cookbooks, they became a standard ingredient in typically English recipes, from "meat and two veg" to carrot cake.[52] Italy did not share in this international success story. At the time when Carotino, followed by his equally goofy companion, the hound Segugio, made his appearance on the pages of Rodari's narrative, carrots remained a commonplace if nutritious root vegetable that had never made itself quite indispensable to Italian cuisine.

The unfulfilled promise of carrots contrasts with the success of strawberries and bilberries, or rather, plants that have assumed their traditional guise. Few twenty-first-century Italian readers of Rodari's novel may realize that this produce, familiar to them from market stalls and billboard publicity, is not the same fruit that their ancestors used to pick in the wild. In the case of strawberries (*fragole*), their emergence as the current commercial crop, valued for sweet flavor and juicy texture, was the result of hybridization of two American species: *Fragaria chiloensis*, domesticated by the Indigenous Mapuche and Picunche peoples of

Chile, and the wild *Fragaria virginiana* from the northeastern United States and southern Canada. What is remarkable about this marriage—an outcome of spontaneous pollination as much as agronomic experiments—is that it took place in mid-eighteenth-century Brittany, where both plants were introduced by the French merchants, soldiers, and naturalists returning from the New World.[53] The resulting cultivar had nothing to do with wood strawberries or *fraises des bois* (*Fragaria vesca* L.), a species native to Europe and characterized by fragrant and delicate but small and dry fruit (Figure 7.4). The situation with bilberries (*mirtilli*) is even more misleading. The fruit sold today under this name in Italian grocery stores is, in reality, blueberries, which come from a North American shrub of the same genus *Vaccinium* and are typically imported from Chile and Peru.

These microhistories not only reveal a widening gap between scientific and everyday taxonomies and nomenclatures but also highlight the confusion even among such seemingly stable categories as native and foreign species. They also draw attention to the mechanisms of introducing plants into the food culture of Italy through assimilation and substitution—as illustrated by the grouping of summer squashes with calabash gourds, the adoption of cornmeal for making polenta, and the use of the word *mirtilli* as an alias for blueberries—without causing tectonic shifts in traditional dietary knowledge. The surprising resilience and adaptability of this age-old wisdom are particularly evident in the case of tomatoes, which are still subjected to a two-phase method of cooking—first, by processing them into canned or jarred paste and, second, by turning this product into an actual pasta sauce—in response to the nightshade's original reputation as unhealthy if not potentially poisonous food.[54]

In the case of the hierarchy of plants, rooted in the doctrine of the Great Chain of Being, it was only natural that, having lost its intellectual credibility in botanical scholarship with the advent of sixteenth-century empiricism, it would take refuge in early modern food culture, where it could be readily mapped onto socially stratified eating habits. What is remarkable about this long-obsolete scheme is that it not only managed to outlive the taxonomies of Cesalpino, Tournefort, and Linnaeus but also found its way into Rodari's novel. This longevity, rather than reflecting the continual belief in the unequal status of plants, was a sign of the already folkloric status of this hierarchy, which put questions of its verifiability or falsifiability beyond the reach of scientific protocols. Its survival, in other words, was a logical outcome of its incorporation into family lore nourished by the thriving market and cooking traditions of Italy, where it joined the fear of drafts, the omens of spilling salt on the table or oil on the floor, and the imperative to eat lentils on New Year's Eve.

FIGURE 7.4
Gherardo Cibo, Wild strawberry (*Fragaria vesca* L.), ca. 1560s. British Library, London / Bridgeman Images.

In conclusion, it remains to be emphasized that botanical knowledge, to be considered traditional or vernacular, does not need to be Indigenous or non-Western. In this regard, Europe, with its long-standing traditions of ancient dietary medicine and medieval herbalist practice, offers a vast potential for future research. What makes its contribution particularly significant is the volume and range of the surviving documentation, both material and textual. This evidence, in addition to botanical illustrations and herbarium specimens, includes annotated herbals, dietary regimens, recipe books, and agricultural treatises, which often bear an imprint of non-elite practices by referencing vernacular names and common uses of plants. The approach of plant humanities, with its close attention to natural history collections and resources, opens new multidisciplinary opportunities for such scholarship, just as it can orchestrate a meeting between Cipollino and Cesalpino, whose paths would have never crossed in the context of more conventional academic work.

Notes

1 I am grateful to Florike Egmond, Allen Grieco, David Gentilcore, and Yota Batsaki for their valuable feedback on this essay.

2 On *Verbascum*, see Xiang Dong, Elijah Mbandi Mkala, Elizabeth Syowai Mutinda, et al., "Taxonomy, Comparative Genomics of Mullein (*Verbascum*, Scrophulariaceae), with Implications for the Evolution of *Verbascum* and Lamiales," *BMC Genomics* 23, no. 566 (2022), https://doi.org/10.1186/s12864-022-08799-9.

3 A popular overview of the commercialization of bananas is Ashley Buchanan, "Bananas: Chaos out of Order," Plant Humanities Lab, https://lab.plant-humanities.org/Banana/.

4 Jürgen Habermas, *The Structural Transformation of the Public Sphere: An Inquiry into a Category of Bourgeois Society*, trans. Thomas Burger, with the assistance of Frederick Lawrence (Cambridge, Mass.: MIT Press, 1991).

5 See Hannah Hardenbergh, "What We Know Now: Tracing the Tamarack," Plant Humanities Lab, https://lab.plant-humanities.org/tamarack/, which draws upon Anne M. Dunn, *Fire in the Village* (Duluth, Minn.: Holy Cow! Press, 2016), primarily addressed to a young audience, as a source of Anishinaabe Indigenous knowledge.

6 Gianni Rodari, *Il romanzo di Cipollino* (Rome: Edizioni di Cultura Social, 1951); in the second edition of 1957, Rodari changed the title to *Le avventure di Cipollino*. The edition cited here is Gianni Rodari, *Le avventure di Cipollino* (Trieste: Einaudi Ragazzi, 2010). On the concept of vernacular knowledge, see Ülo Valk, "An Introduction to Vernacular Knowledge," in *Vernacular Knowledge: Contesting Authority, Expressing Beliefs*, ed. Ülo Valk and Marion Bowman (Sheffield: Equinox, 2022), 1–21.

7 Rodari's authoritative biography is Marcello Argilli, *Gianni Rodari: Una biografia* (Turin: Einaudi, 1990).

8 Rodari, *Le avventure di Cipollino*, 17.

9 Rodari, *Le avventure di Cipollino*, 30, 56–57, 65, 80.

10 Rodari, *Le avventure di Cipollino*, 105–9.

11 Rodari, *Le avventure di Cipollino*, 57–58.

12 Rodari, *Le avventure di Cipollino*, 195–200.

13 Rodari, *Le avventure di Cipollino*, 151.

14 Rodari, *Le avventure di Cipollino*, 155.

15 Rodari, *Le avventure di Cipollino*, 7.

16 Giovanni Battista della Porta, *De humana physiognomonia*... (Vico Equense: Giuseppe Cacchi, 1586); the same "physiognomic" method was also applied to plants in Giovanni Battista della Porta, *Phytognomonica*... (Naples: Orazio Salviano, 1588).

17 Allen J. Grieco, "The Social Politics of Pre-Linnaean Botanical Classification," *I Tatti Studies in the Italian Renaissance* 4 (1991): 131–49.

18 Theophrastus, *Historia plantarum* I.iii.1.

19 Andrea Cesalpino, *De plantis libri XVI* (Florence: Giorgio Marescotti, 1583).

20 Carl Linnaeus, *Classes plantarum*... (Leiden: Conrad Wishoff, 1738), 1–32.

21 Quentin Hiernaux and Corentin Tresnie, *Andrea Cesalpino's De Plantis Libri XVI (1583) and the Transformation of Medical Botany in the 16th Century: Edition, Translation, and Commentary on Book I* (Berlin: De Gruyter, 2023), 142–65.

22 On the Venus flytrap, see John R. Schaefer, "Venus Flytrap: Queen of the Carnivorous Plants," Plant Humanities Lab, https://lab.plant-humanities.org/venus_flytrap/.

23 Théodore Caruel, *Illustratio in hortum siccum Andreae Caesalpini* (Florence: [Le Monnier], 1858), 4 (translation by the author). This passage comes from Cesalpino's Italian letter (September 14, 1563) that accompanied his herbarium addressed to Alfonso Tornabuoni, Bishop of Borgo Sansepolcro, which is now in the Museum of Natural History of the University of Florence.

24 Anatole Tchikine, "Environmental and Natural Science: Early Modern Botanical Knowledge," in *A Cultural History of Nature in the Early Modern Era*, ed. Luke Morgan (London: Bloomsbury, forthcoming).

25 On the Lyon herbal, its regional context, and authorship, see Oury Goldman, "Produire l'universalité savante depuis l'arène locale: Les cartographies végétales de l'*Historia generalis plantarum* (Lyon, 1586)," *Bulletin de l'Association des historiens modernistes des universités françaises* 45 (2024): 34–53.

26 *Historia generalis plantarum*... (Lyon: Guillaume Rouillé, 1586), n.p. ("Index librorum historiae generalis plantarum") (translation by the author).

27 Michel Foucault, *The Order of Things: An Archaeology of the Human Sciences* (London: Routledge, 1986), xv–xix. Foucault's comments leave no doubt as to his conviction that, whether based on an actual Chinese source, the "encyclopedia" reflected non-Western intellectual attitudes. Borges's knowledge of China, however, was largely limited to his interest in Chinese classic fiction, which he read in English and German translations, suggesting that the animal catalogue was likely his own invention; see Rosario Hubert, "Sinology on the Edge: Borges's Reviews of Chinese Literature (1937–1942)," *Variaciones Borges* 39 (2015): 81–101. I am grateful to Lucas Mertehikian for bringing this article to my attention.

28 For a critique of Foucault's historical epistemology, see Ian Maclean, "Foucault's Renaissance Episteme Reassessed: An Aristotelian Counterblast," *Journal of the History of Ideas* 59, no. 1 (1998): 149–66.

29 Foucault refined the concept of *heterotopia* in his essay "Des espaces autres" (1984), based on a lecture given in 1969; the title is variably translated as "Of Other Spaces" or "Different Spaces." See Michel Foucault, "Of Other Spaces," trans. Jay Miskowieç, *Diacritics* 16, no. 1 (1986): 22–27; and Michel Foucault, *Aesthetics, Method, and Epistemology*, ed. James D. Faubion, trans. Robert Hurley et al. (New York: The New Press, 1998), 175–85.

30 Tchikine, "Environmental and Natural Science."

31 The concept of the "period eye" was introduced in Michael Baxandall, *Painting and Experience in Fifteenth-Century Italy* (Oxford: Oxford University Press, 1972), but has important applications beyond the field of art history.

32 Agostino del Riccio, *Agricoltura sperimentale*, vol. 56, I, fol. 74v, Targioni Tozzetti, Biblioteca Nazionale Centrale, Florence.

33 Agostino del Riccio, *Agricoltura sperimentale*, vol. 56, II, fol. 382r, Targioni Tozzetti, Biblioteca Nazionale Centrale, Florence.

34 On the sixteenth-century interest in cypress vine, see Giovan Vettorio Soderini, "Il trattato della cultura degli orti e giardini," in *Le opere*, ed. Alberto Bacchi Della Lega (Bologna: Romagnoli Dall'Acqua, 1902–1907), 2:284–85.

35 Ken Albala, *Eating Right in the Renaissance* (Berkeley: University of California Press, 2002), 10, 12–13, 96, 181, 206, 242–43, 254–55 (especially n37), 267–68.

36 Grieco, "The Social Politics of Pre-Linnaean Botanical Classification," 132–33.

37 Plautus, *Mostellaria* 39; Plautus, *Poenulus* 1314; Persius, *Satires* IV.35–38; and Martial, *Epigrams* XII.32.17–18. For the social connotations of garlic, see Emily Gowers, *The Loaded Table: Representations of Food in Roman Literature* (Oxford: Clarendon Press, 1993), 290–92.

38 Cited in Robert W. Gaston, "Peeling the Onion: Experiencing the Senses in Bronzino's Burlesque Poem *La cipolla*," in *Sense and the Senses in Early Modern Art and Cultural Practice*, ed. Alice E. Sanger and Siv Tove Kulbrandstad Walker (Farnham: Ashgate, 2012), 98.

39 Gary Paul Nabhan, *Songbirds, Truffles, and Wolves: An American Naturalist in Italy* (New York: Pantheon Books, 1993), 74–75.

40 David Gentilcore, *Pomodoro! A History of the Tomato in Italy* (New York: Columbia University Press, 2010), 45–98.

41 Rodari, *Le avventure di Cipollino*, 81, 152, 155, 159.

42 Teresa A. Lust and Harry S. Paris, "Italian Horticultural and Culinary Records of Summer Squash (*Cucurbita pepo*, Cucurbitaceae) and Emergence of the Zucchini in 19th-Century Milan," *Annals of Botany* 118, no. 1 (2016): 53–69, especially 66.

43 Rodari, *Le avventure di Cipollino*, 82.

44 Soderini, *Le opere*, 2:400.

45 Antonino Venuto, *D'agricoltura, nel quale si insegna il vero modo di coltivare i campi, i prati, gli orti, i giardini, le viti, gli arbori, & tutte le cose utile & necessarie, che s'appartengono a l'huomo in materia di villa* (Venice: Francesco Rampazetto, 1550), 4.

46 Giovanni Gioviano Pontano, *Eclogues: Garden of the Hesperides*, ed. and trans. Luke Roman (Cambridge, Mass.: Harvard University Press, 2022).

47 Giovanni Battista Ferrari, *Hesperides sive de malorum aureorum cultura et usu* (Rome: Hermann Scheus, 1646).

48 Anatole Tchikine, "Beyond the Leges Hortorum: Gardens in the Urban Landscape of Early Modern Florence," *Journal of the Society of Architectural Historians* 83, no. 4 (2024): 437–38.

49 Rodari, *Le avventure di Cipollino*, 156, 161.

50 Soderini, *Le opere*, 2:103–4; and Giacomo Castelvetro, *The Fruit, Herbs and Vegetables of Italy: An Offering to Lucy, Countess of Bedford*, trans. and ed. Gillian Riley (London: British Museum, 1989), 138.

51 Giovanni Miniati da Prato, *Narrazione, e disegno della terra di Prato di Toscana* (Florence: Francesco Tosi, 1594), 50–52.

52 John Stolarczyk and Jules Janick, "Carrot: History and Iconography," *Chronica Horticulturae* 51, no. 2 (2011): 17.

53 James F. Hancock, *Strawberries*, 2nd ed. (Boston: CAB International, 2020), 32–58.

54 I am grateful to David Gentilcore for sharing this observation with me; see also Gentilcore, *Pomodoro*, ix.

8

HERBARIUM AS LIVING ARCHIVE

ASHANTI SHIH

A few years ago, I was digging through herbarium specimens at the Bernice Pauahi Bishop Museum in Honolulu, Hawaiʻi. While paging through the dried plant specimens, I came across a specimen of ʻeke silversword (*Argyroxiphium caliginis*), a rare flowering plant endemic to the island of Maui and belonging to the sunflower family. In the late 1920s, two men collected this plant and transformed it into a scientific specimen: Henry Wiebke, a Kanaka Maoli (Native Hawaiian)[1] botanist and medical student; and Otto Degener, a German American botanist who had recently settled in Hawaiʻi from New York. From Degener's papers and my correspondence with Wiebke's granddaughter, I knew that the men maintained a close collecting partnership for six years, creating thousands of specimens and working on two books together.[2] I had already seen hundreds of Degener and Wiebke's jointly collected specimens at the New York Botanical Garden, and I knew that there were thousands more to browse through at the Bishop Museum and other herbaria across the United States.

However, in Honolulu that day, this herbarium specimen stood out to me. It is a highly valued specimen: a type specimen of an extremely rare plant, endemic to and existing in only a single volcanic crater on Maui. I had intentionally asked to pull these specimens due to the botanical significance of the ʻeke silversword. But there was an unexpected detail on the specimen's label (Figure 8.1). Wiebke's name had been typed next to Degener's—to indicate that both men had collected and prepared the specimen together—and then, it seemed, Wiebke's name was later crossed out. Crossed-out names are not uncommon on herbarium specimens, which are full of idiosyncrasies despite their otherwise standardized form. Yet Wiebke's name had clearly been intentionally written in and then intentionally negated.

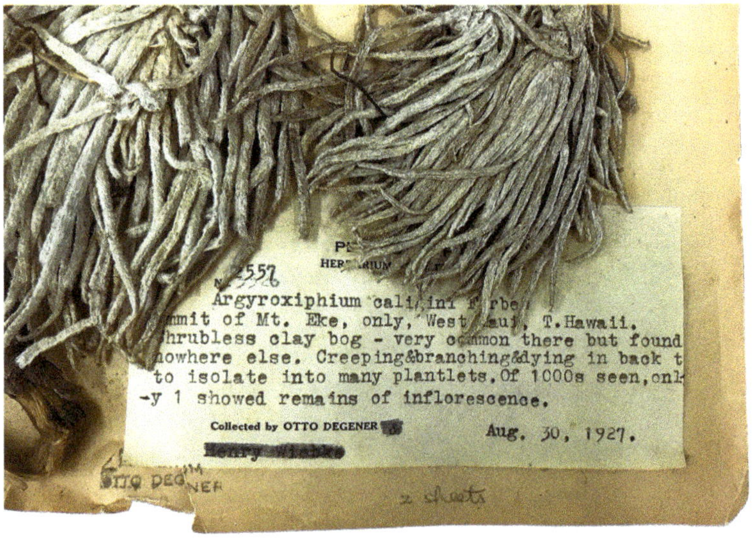

FIGURE 8.1

The label of an herbarium specimen of *Argyroxiphium caligini* Forbes that was collected by Otto Degener and Henry Wiebke on Mt. Eke, Maui, in 1927. Wiebke's name was added with a typewriter and then crossed out. Specimen housed at the Bishop Museum Herbarium Pacificum, Honolulu. Photograph by Ashanti Shih.

While nothing could be proven, the striking of Wiebke's name, from this and from dozens of other specimens I viewed that day, aligns with the fact that the two men eventually had a falling out over issues of authorship, money, and status, with Degener denying Wiebke coauthor credit on their two books. One of these books, *Flora Hawaiiensis*, was issued in multiple volumes and remains an authoritative text on Hawaiian flora.[3] With no mention of Wiebke, the first volumes are instead credited solely to Otto Degener, and later volumes to both him and to botanist Isa Degener (née Hansen), whom Degener married later in life and who is sometimes misleadingly called his "lifelong collaborator," despite their partnership starting decades after that of Wiebke and Degener.[4]

Moreover, Wiebke's lack of credit (and Degener's sole credit) has been continually reproduced beyond these simple strike-outs on individual specimens and beyond *Flora Hawaiiensis*. Degener, sans Wiebke, was credited for the ʻeke silversword specimens in the Bishop Museum herbarium's digital catalogue; and similar specimens and records exist in other herbaria, such as that at Harvard University.[5] Whereas Wiebke has been entirely forgotten by those in the botanical world, Otto Degener, in both his lifetime and his posthumous legacy, has been an outsized figure in taxonomic botany in Hawaiʻi, nationally, and across the globe. Herbaria in Hawaiʻi and North America house Wiebke's specimens but do not know his biography or the fact that he was Kanaka Maoli, due to his Anglo-German name from his father's side (Figure 8.2). In contrast, Degener is considered the most prolific collector of native Hawaiian flora ever, with the largest and finest collection of Hawaiian Island herbarium specimens in existence.[6] Degener was also an avid collector in Fiji and Bermuda. One of his most significant contributions to taxonomic botany was finding the missing link between angiosperms and gymnosperms in a Fijian tree (*Degeneria vitiensis*) and establishing from this a new family of angiosperms (Degeneriaceae). This contribution, in addition to numerous publications and his vast specimen collection, brought local, national, and international acclaim, including a Linnaeus Medal from the Royal Swedish Academy of Sciences. Over the sixty years that he spent living and working in Hawaiʻi, Degener boasted a career-long affiliation with the New York Botanical Garden, and he held prestigious positions around the islands, including serving as the first park naturalist of Hawaiʻi National Park, a botany professor at the University of Hawaiʻi, and a collaborator with the Bishop Museum. Moreover, he is remembered in Hawaiʻi as an early and passionate advocate for

FIGURE 8.2
Henry Wiebke with his daughters, ca. 1926. Retrieved from "Henry Wiebke," *FamilySearch*, https://familysearch.org/tree/person/about/KW84-FBM.

native species restoration, alien species mitigation, and the expansion of nature preserves.[7]

Degener's outsized accomplishments and Wiebke's complete lack of professional legacy reflect Wiebke's erasure as a marginalized knowledge producer. Put in conversation with other oral and archival sources, this striking of Wiebke's name from the record should be read as an act of Native erasure and a settler claim to authorship. This is a familiar story in histories of colonial science and collecting, in which a colonial expert claims authority by exploiting marginalized peoples' knowledge, labor, and networks without credit or compensation. Scholars have shown how, across vast geographies, temporalities, and fields, this "invisible labor" underwrites the production of scientific knowledge and that the privilege of authorship—including the benefits of reputation and profit—has largely been reserved for an elite group of white, upper-class men.[8]

Thousands of other specimens categorized in the "Herbarium of Otto Degener" present yet another dimension to this story, which further speaks to the invisible labor that goes into creating colonial knowledge and to how that knowledge is archived and reproduced today. Namely, in the early twentieth century, in the U.S. plantation colony of Hawai'i, Degener was not only collaborating with Wiebke and other Kanaka Maoli knowers but also with dozens of Asian men who collected and illustrated plants with him. Their very presence, and their positionality as racialized actors whose migrations stem from global routes of Asian indentured (or "coolie") labor, is evidenced by their names appearing on thousands of specimen labels and in their handwritten notes about the plants. Degener's herbarium represents the contributions of Degener, Wiebke, and dozens of other white settler, Kanaka Maoli, Japanese, Chinese, Korean, and Filipino collaborators who worked with Degener before he married in the early 1950s. While holding a radically different relationship to Indigenous land and knowledge than Kanaka Maoli men like Wiebke, these Asian collaborators also experienced marginalization. They were treated as replaceable laborers and their work was subsumed under Degener's name and reputation. Similar to how the institution and political economy of slavery shaped knowledge production in the Caribbean, Hawai'i's settler colonial plantation society fundamentally shaped the way that botanical knowledge was produced in Hawai'i.[9] How should we read these erasures and betrayals in the historical process of producing colonial scientific knowledge, and in the ways that it is archived and interpreted today? How should humanists engage with science archives like herbaria, which were created in colonial contexts and are still implicated in colonial legacies, including reproducing and extending the erasure of Indigenous and racialized collaborators?

In this chapter, I use the specimens that this expansive group of men collected to explore the potential of the herbarium as a site for inquiry in the plant humanities. Historians of science have emphasized that herbarium specimens, like other scientific specimens, are static entities that have been transformed from dynamic living things into dead matter, stripped of ecological and social contexts and placed on a Linnaean sheet, or what Daniela Bleichmar calls the "white space."[10] These specimens are material evidence of Western scientific approaches, including standardization, universality, and immutable mobility. When an individual plant is collected, dried, and pasted to a sheet according to Linnaean standards, it transforms from a specific individual into a universalizable representative of a *species*; this specimen can then travel across time and space to be studied and referenced.[11] The herbarium specimens that Wiebke, Degener, and their Asian collaborators collected undoubtedly follow this pattern. The fact that they were taking Hawaiian plants and removing them from their native contexts, including Indigenous lands and knowledge systems, to turn into dead specimens is significant for thinking through how Western colonial botany has helped erase marginalized collaborators and their respective relationalities to nonhuman nature to serve U.S. settler colonial claims in Hawai'i.

While recognizing the death and erasure that the immutable mobile brings, might we also recover these specimens, and the herbarium more generally, as a *living* archive? Looking beyond the obvious frozen temporality and decontextualization of the herbarium, how can we breathe life back into herbarium specimens? How might we use herbaria to recover marginalized perspectives and open up new possibilities for addressing present colonial legacies and imagining decolonial futures? Feminist and postcolonial scholars have argued for the importance of reading colonial archives not just as spaces of silencing and erasure but also as spaces of dynamism and radical possibilities, where archival materials can be read against the grain or layered and crosscut according to subaltern modes of inquiry. As Black feminist scholar Marisa J. Fuentes writes, reading an archive for alternative meanings is like "where fabric is cut at an angle to produce elasticity."[12] How can we read herbarium specimens—and perhaps science archives more generally—for elasticity, dynamism, and possibilities? "Stretching the archive," in Fuentes's words, can help us recover marginalized actors who participated in Western botany and reveal the dynamic social and political contexts of colonial scientific knowledge production. We might rethink herbarium specimens as records of interactions between Indigenous, diasporic, and Western actors and their perspectives, including their respective knowledge systems and relationalities to the nonhuman world. Finally, finding elasticity in the herbarium also means recognizing the archive itself as an evolving creature that plays an ongoing role in Indigenous erasure, colonial legacies, and possible imagined futures. As Kathleen Cruz Gutierrez writes, "working against the endurance of white space demands a recovery of the contexts in which species were discovered, collected, and described," and this act can help us see the "afterlives of imperial war and extraction" that colonial botany left in its wake.[13] Bringing a humanistic lens to what scientists already recognize as a "vitality" of annotative practices on herbarium specimens,[14] how might we stretch the "white space" to read them for past lives and present hauntings?

Reading the Herbarium for Diasporic Knowledges and Plantation Colonialism

Reading herbarium specimens against the grain, as a living record, allows for the recovery of marginalized actors who participated in Western botany and can reveal the dynamic social and political contexts of colonial scientific knowledge production. Herbarium specimens, in this way, can be said to be a record of the political and social worlds in which the plants were collected. Herbarium specimens from early twentieth-century Hawaiʻi reveal that collection occurred in the contexts of the plantation, which fed on Asian migrant labor, and the settler colony, built on Kanaka Maoli land.

Traces of the plantation system, heavily reliant on Asian migrant labor, exist everywhere in Degener's practice and in his herbarium collection. Degener built his botanical practice in the economic, political, and social contexts of the plantation. Upon his arrival in Hawaiʻi in the 1920s, he secured a teaching position at the University of Hawaiʻi and quickly began to expand his collecting practice by employing assistants and illustrators. He explained to a friend in New York: "I need collectors, often trail cutters, someone to press, p[oison] and mount my specimens, pack horses, and above all, student 'artists' to take into the field."[15] To fill these positions, Degener took advantage of the labor pool that surrounded him. He hired dozens of young Asian men, many of whom had either migrated to Hawaiʻi to work on the plantations themselves or were second- or third-generation descendants of plantation workers. Most of these Japanese, Chinese, Korean, and Filipino men encountered Degener in his university botany classes and in public spaces that catered to white settlers, like hotels and restaurants. Others, like Korean high schooler Kwan Kee Park, met Degener through referral.[16] These men worked and lived with Degener, and at times also with Wiebke, for months or years at a time. Together, they traversed Oʻahu and embarked on expeditions around the Hawaiian Islands to collect native flora. They turned their findings into herbarium specimens and botanical illustrations, with the Asian collaborators performing most of the illustration work in addition to collecting specimens out in the field.

One of Degener's longest-term collaborators was Kwan Kee Park, who was just sixteen years old when he began working with Degener in the late 1920s. Park had immigrated to Hawaiʻi from Korea with his family in 1923, a year before Asian

FIGURE 8.3
Botanical illustrator Kwan Kee Park at Otto Degener's home in Honolulu, ca. late 1920s or early 1930s. Otto Degener Records, Archives of the New York Botanical Garden, New York.

immigration to the United States and its territories would be restricted to the point of effective exclusion.[17] One of Park's high school teachers, who knew Park by his English name "James," took note of Park's interest in painting and drawing and introduced him to an acquaintance, American botanist Otto Degener, who was working locally to collect and document Hawaiian plants.[18] Park pedaled his bicycle to Degener's house in Honolulu in the afternoons to draw and collect with Degener until, upon high school graduation, Park moved into one of Degener's spare rooms, on a side of the house that Degener referred to as his "servants' quarters" (Figure 8.3).[19] There, Park lived with a few of Degener's other collaborators, a rotating group of young Asian and Kanaka Maoli men who cleaned, drove, collected, and illustrated for Degener. In return, Degener provided boarding, paid a modest wage, and issued loans to the men who wished to pursue further schooling. Park's collaboration with Degener even took him and a young Filipino man named Tony Agliam to New York, where they helped Degener deposit forty thousand herbarium specimens in the New York Botanical Garden's collection and work on the manuscript of *Flora Hawaiiensis*.[20]

Agliam and Park are illustrative of the various kinds of Asian migrants who chose to collaborate with Degener: Agliam had originally immigrated to Hawai'i to work on a sugar plantation, while Park had no recorded connections to plantation work. Their differences reflect the changing class status and expectations of some Asian ethnic groups in Hawai'i during that time. By the time of Degener's arrival in Hawai'i, federal exclusion laws prohibited Chinese, Japanese, and Korean laborers from immigrating. Planters turned to recruiting labor from the Philippines, since Filipinos, as U.S. colonial subjects, could travel throughout the United States and its territories.[21] Many of the Filipino men who worked with Degener were, therefore, first-generation immigrants to Hawai'i and were directly recruited to work on the plantations.[22] Even after fulfilling their plantation contracts, Agliam and other Filipino men, like Valentino Arlantico, alternated between illustrating for Degener and working in the fields and canneries.[23] With the exception of Kwan Kee Park, Degener's other artists were of Japanese or Chinese descent. Most were already a generation or two removed from the plantations and were moving up in class by attending the university to study agricultural science or medicine. These aspirational pursuits echoed the racial hierarchies of the plantation, where Japanese and Chinese workers were increasingly trained for indoor, technical work, while Filipinos, residing at the bottom of the plantation's racial hierarchy, were relegated to the fields with little hope for advancement.[24]

Despite Park's relative distance from plantation labor, his herbarium specimens and others in Degener's collection reveal that Park and other Asian collaborators navigated the social, economic, and political reality of the plantation system daily. For instance, take one of Park and Degener's specimens of yomogi (*Artemisia princeps* Pampan or Japanese mugwort) from 1933. While hiking along a trail in Kahuku, Oʻahu, the two men picked the yomogi plant, a common East Asian herb.[25] Cutting, pressing, and later pasting it to their Linnaean sheet, Park and Degener wrote on the specimen label that the yomogi had been found "[p]lanted and escaping at Japanese laborers' pineapple camp; said to be used as tea" (Figure 8.4). This short description opens myriad windows into the world in which they were collecting plants. First, there is the striking reference to pineapples. The fruit was not native to Hawaiʻi but was a plantation introduction and the second most important export crop in Hawaiʻi apart from "king" sugar.[26] In nearby Wahiawa, industrialist James Dole—cousin to Stanford B. Dole, who led the overthrow of the Kanaka Maoli monarchy and served as the first president of the white settler government—was running pineapple plantations and canneries, exporting and introducing the fruit to white middle- and upper-class consumers in the United States.[27] Other specimens in Degener's herbaria reveal that he and his collaborators collected plants at or near plantations: *Emilia fosbergii* is described on a label as a "pineapple field weed" in Koloa, Kauai, and one *Ipomoea triloba* specimen was found in a "pineapple field" near Kapapa Gulch, Oʻahu.[28]

Second, Park and Degener described the yomogi as having been "planted" at the "camp"

FIGURE 8.4

The 1933 label on Kwan Kee Park and Otto Degener's specimen of *Artemisia princeps* Pampan (originally classified as *Artemisia vulgaris* L.) alludes to the plots that Japanese migrant laborers planted in the shadow of Hawaiʻi's plantations. Park and Degener note that they found this yomogi (Japanese mugwort) "[p]lanted and escaping at Japanese laborers' pineapple camp; said to be used as tea." Bishop Museum Herbarium Pacificum, Honolulu.

of "Japanese laborers," which gestures toward the racialized workforce of the plantations. It also suggests that, perhaps, the yomogi had been intentionally cultivated by Japanese migrants, apparently for tea leaves, and perhaps also in a kind of "plot" or subsistence garden in the shadow of the plantation.[29] We can see this in other specimens from Degener's herbarium in the 1920s and 1930s: a *Gossypium barbadense* (a tropical type of cotton) plucked from a "Chinese garden" near Kawela Bay on Oʻahu, or a specimen of *Ipomoea aquatica* (water spinach or ong choy) found in a "rice field" in Kailua, Oʻahu.[30]

Finally, by writing that the plant was "escaping" the camp, Park and Degener recognize the agency of the yomogi itself, which, though introduced and cultivated by human hands, was growing according to its own needs and desires. Through this simple note, Park and Degener's herbarium specimen divulges the economic, social, and political worlds of 1930s Hawaiʻi.

Beyond finding plants in the plantation landscape, Degener's use and characterization of Asian labor entangled their collecting practice in the plantation economy and its racial hierarchies. Degener clearly associated his Asian collaborators with the plantations and the stereotype of them as the ideal laborer, the "coolie," who would be quiet, docile, and hardworking. He openly used plantation wages as the standard for his Asian assistants' meager pay, which demonstrated their status as racialized wage laborers within the larger plantation economy, despite performing scientific work. In the mid-1920s, Degener was paying about forty cents per hour for freelance illustration and collecting work. For the men who lived with him, the pay was much less—one dollar per day for untracked hours of scientific and domestic work. Degener felt this situation was reasonable considering he was also providing them room and loans for their education. He later defended the meager payment, writing: "Remember that sugar plantation laborers, with free housing & medical care but nothing else, were being paid $1.00 per day for arduous labor in the fields. I paid my boys $1.00 per day and keep[,] as well as continued their schooling, a most satisfactory wage."[31] Degener may have believed that his arrangement differentiated him from planters, but his supposed patronage mirrored the paternalistic practices that planters developed during this period to quell labor unrest. Like Hawaiʻi's white planter elites, Degener was promising benefits like limited training and lodging but only to maintain hierarchy and loyalty.[32] Moreover, by performing domestic service for Degener, the Asian men who lived or traveled with him were further debased to the status of "houseboys"—a common mode of feminized work for young Asian men in white U.S. households during this period.[33] The local contexts of the plantations, in the era of Asian exclusion in the United States more generally, gave form to the ways that Asian settlers were able to collaborate with Degener and to pursue botanical work as a limited form of socioeconomic mobility. The thousands of specimens housed at the Bishop Museum, Harvard University, and the New York Botanical Garden that bear the names of Kwan Kee Park, Tony Agliam, and dozens of other Asian men remind us that their labor and the broader context of the plantation were essential to building Degener's reputation during his lifetime and in his posthumous legacy as one of the greatest collectors of Hawaiian flora.

Reading the Herbarium for Indigenous Knowledge and Colonial Haunting

Beyond recapturing the political, economic, and social milieu of the plantation system, we might rethink herbarium specimens as traces of the interactions between Indigenous, diasporic, and Western actors and their perspectives, including their respective knowledge systems and relationalities to the nonhuman world. Just as the yomogi cultivated in Japanese laborers' camps provides a glimpse of Asian diasporic relationships to plants and the plantation, other specimens in Degener's herbarium reveal Indigenous cultivations and cosmologies of plants. Mentions of "native gardens" frequently occur on the specimen labels that Degener and his collaborators collected across the Hawaiian Islands, especially those collected from the more developed island of Oʻahu. In 1933, Degener collected an *Artemisia princeps* Pampan (yomogi or Japanese mugwort) and wrote on the label that he had found the specimen "[p]lanted in native's garden" in Kawela Bay on O'ahu (Figure 8.5).[34] Clearly, Kānaka Maoli lived in close relation to plants, both native and introduced, and actively cultivated them in domestic spaces.

Beyond the literal cultivation of plants by Indigenous actors, we might also apply Fuentes's idea of "elasticity" to these specimens, to read them for something more ambiguous but nonetheless ever-present: Indigenous cosmologies and the haunting effect of settler colonialism in the islands. Kānaka Maoli, like many other Indigenous peoples, have long-standing relationalities to their lands, plants, and other forms of life and landscape, which are often defined in notions of kinship, animacy, and reciprocal care. Kanaka Maoli scholars, like Lilikalā Kame'eleihiwa, emphasize that Kanaka Maoli identity, language, and cosmologies are fundamentally defined in relation to land and nature. This is clear in Kanaka Maoli oral traditions (mo'olelo), including the origin story, the Kumulipo. In this story, the first ancestor, Hāloa, was born from the kalo (the taro plant), who is considered the source of life, the source of Hāloa's name, and the elder sibling of Kānaka Maoli.[35] We might read the native gardens as evidence of reciprocal care.

Herbarium specimens also testify to the fact that Degener was practicing on Indigenous land, with Indigenous history and ongoing Indigenous presences. In 1923, Degener recorded that he had found a *Blumea laciniata* in a "clearing at abandoned native settlement" near an important site in Kanaka Maoli history: Nuuanu Pali on O'ahu.[36] That same year, he collected a specimen of *Cucumis dipsaceus* at Makapu'u Point, a "near abandoned native city on way to Waimanalo, Oahu."[37] A few years later, on an expedition to Moloka'i, Degener wrote on a label of *Dioscorea alata* L. (uhi or purple yam) that he had found the plant in Wailau Valley, "growing just mauka of native shacks" (Figure 8.6).[38] Using the Kanaka Maoli directional word "mauka" to mean "toward the mountain side," rather than Western directional terms like north or east, Degener was both operating within an island-oriented, Native Hawaiian geography and also noting an ongoing Indigenous presence.[39] The curious attachment of a page of notes about the Native cultivation of yams, an important staple crop that Polynesians originally brought to Hawai'i on outrigger canoes, further reveals Degener's interest in the roles that plants played in Kanaka Maoli worlds.

FIGURE 8.5
Kanaka Maoli presence and cultivation are apparent in Degener's 1933 label of *Artemisia princeps* Pampan (originally *Artemisia vulgaris* L.), which he found "[p]lanted in native's garden" in Kawela Bay. Bishop Museum Herbarium Pacificum, Honolulu.

Degener's immersion in Kanaka Maoli cosmology took place through the help of Indigenous collaborators such as Henry Wiebke and Oliver

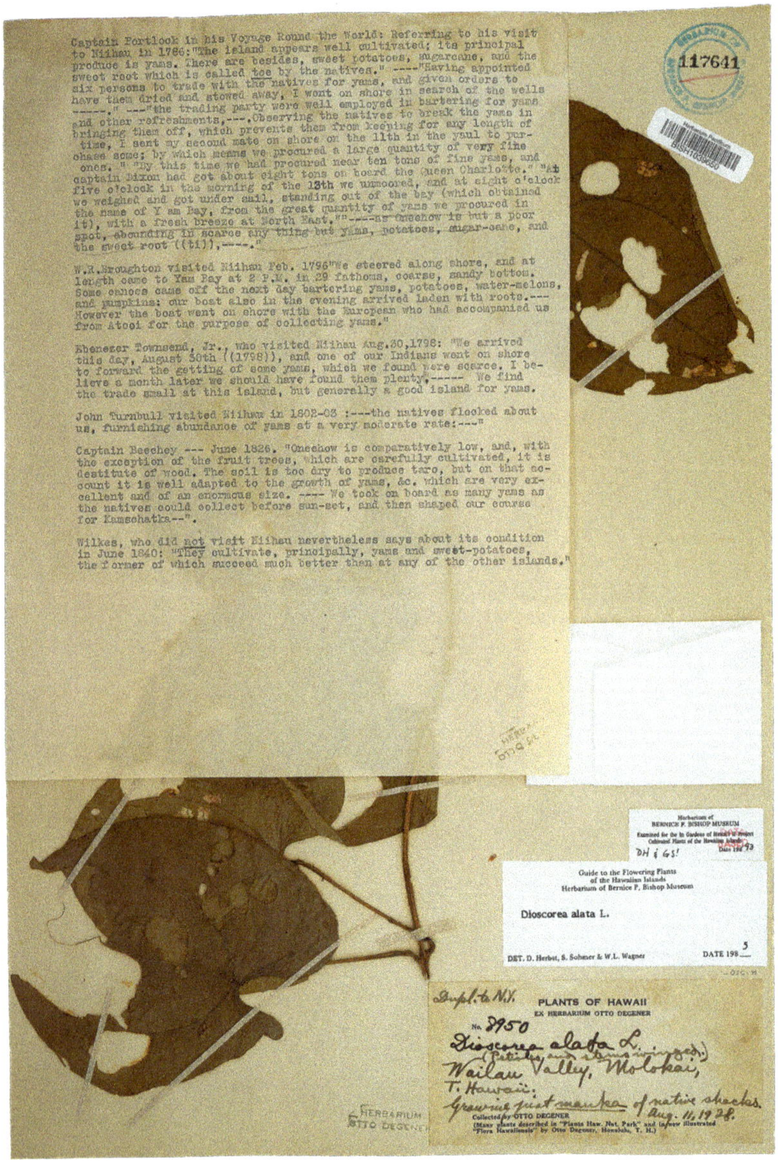

FIGURE 8.6

Degener's 1928 label on this *Dioscorea alata* L. (uhi, or purple yam) specimen claims that he found the plant in Wailau Valley, "growing just mauka of native shacks." The Kanaka Maoli directional word "mauka" denotes that the yam was growing "toward the mountain side" and reveals an Indigenous cosmology. Atop the specimen sits a page of notes, presumably for one of Degener's ethnobotanical publications, explaining how Kānaka Maoli historically cultivated yams as a staple crop. Bishop Museum Herbarium Pacificum, Honolulu.

Pohina. Wiebke, as a medical student and botany laboratory assistant at the University of Hawai'i, was trained in Western botany. Collecting in partnership with Degener over six years, Wiebke even discovered a plant in the aster family, a beggarticks endemic to Moloka'i, and named it after himself: *Bidens wiebkei*. However, in his botanical insights, Wiebke drew on his knowledge of Indigenous cosmology, language, and people—and Degener relied greatly on Wiebke for such knowledge and access. Wiebke consistently discerned the Hawaiian uses and names of the plants that he and Degener were collecting to determine their significance to Western botany. In one case, Degener asked Wiebke about the Hawaiian name for the red fern *Sadleria*. Wiebke answered: "Amaumau is the name of the smaller Sadleria—the larger one is called Amau ['ama'u]," and he explained several uses for 'ama'u, including as a material for thatching huts, a ground cover for temples, and a dye.[40] Wiebke noted that Hawaiians traditionally covered the ground with 'ama'u leaves when planting vegetables. When it was about to rain, they would remove the 'ama'u leaves to plant the seeds and then replace the leaves after the rain to prevent evaporation. "There is a saying," Wiebke wrote: "While rain clouds are still in heaven remove the covering of amau leaves." After disclosing this abundance of information, Wiebke told Degener to follow up with other Kānaka Maoli, who lived in closer relation with the fern than Wiebke. "Better check up with Hawaiians down there who see the plant daily," he told Degener.[41] Well-versed in both Kanaka Maoli knowledge and Western botany, Wiebke thus acted as a broker between the two, melding them together to form a crucial component of his and Degener's knowledge production.

In addition to Wiebke's expertise, Degener also drew on the knowledge and connections of another young Kanaka Maoli man, Oliver Pohina. Pohina was a fisherman from Miloli'i, South Kona, Hawai'i Island.[42] While it is unclear how Degener and Pohina first met, the two exchanged several letters in the late 1920s and early 1930s that worked to clarify Degener's understanding of Hawaiian plants. Like Wiebke, Pohina was tapping into the knowledge that he and other Kānaka Maoli possessed to help Degener correctly identify plants.

After Degener sent him a list of plant descriptions with accompanying names, Pohina informed Degener that everything was right except for the plant labeled "Kukaenene." Unsure of the plant's name himself, Pohina had asked a friend and then wrote Degener, "a Hawaiian told me that, that is not the name for that plant but a different name and he has forgotten it."[43] Talking with other Hawaiians in Kona, Pohina also gathered stories about other plants like iliahi, wrote them down, and sent them to Degener: "You can read it over and see if it is use full [*sic*] to your book."[44] Clearly, Kanaka Maoli collaborators like Wiebke and Pohina intimately knew these plants, their uses, and their ecological relationships, or knew of Indigenous knowers and growers who did. Moreover, the myriad uses and continued relationalities that Wiebke and Pohina uncovered presented the plants as fundamentally familiar and entangled in an active Indigenous culture practiced by living peoples.

Moreover, Degener would frequently send Pohina physical plants in the mail to ask him to help identify the plants' Hawaiian names and uses.[45] As with Degener's confusion between ʻamaʻu and ʻamaʻumaʻu in his conversations with Wiebke, Degener's exchanges with Pohina reveal the botanist's struggles with Hawaiian spellings and his inability to find and identify plants without the help of Native informants. In one instance, Degener sent Pohina some scraps of what he believed to be "two common kinds of Kokolau," and asked him if they matched the plants that Pohina told Degener he had seen in the wild: "Is this one of the plants you say is growing near Milolii or is it different? If it has <u>big</u> yellow petals it is a native kind which I would like to get, some day." Pohina wrote back, correcting Degener's spelling and classification: "The plants that you send over is not a Kookoolau[.] [T]he one with the yellow flower is called nehe and the other one is called moonihoawa but used as tea. The kookoolau here does not grow any more[.] [T]he cattles and goats have eaten them all up but I have heard it[']s along the road at Puuwaawaa."[46] Once again demonstrating his Indigenous knowledge and knowledge networks, Pohina was helping Degener define the *Bidens* genus, a group of various subspecies which are collectively called koʻokoʻolau in ʻŌlelo Hawaiʻi (the Native Hawaiian language) and the same genus to which Wiebke's discovery of a year prior, *Bidens wiebkei*, belonged.

Pohina went beyond simply correcting Degener's naming and siting of the genus, however. He also helped Degener collect various subspecies of koʻokoʻolau when Degener visited Hawaiʻi Island the following spring. Several herbarium specimens from a week in March 1930 reveal that Pohina and Yasumi Iwasaki, one of Degener's frequent Japanese collaborators, traveled along the Kohala coast together, weaving in and out of valleys at Pololū and Niulii to collect a variety of native plants. Pohina and Iwasaki collected several subspecies of koʻokoʻolau, including *Bidens hillebrandiana* subsp. *hillebrandiana* and *Bidens hawaiensis* (Figure 8.7).[47] These specimens and their absorption into Degener's herbarium demonstrate Degener's reliance on Indigenous knowledge, networks, and language to translate native Hawaiian plants into scientific specimens with taxonomic significance. That Iwasaki was involved echoes, again, Degener's dual reliance on Indigenous informants and Asian labor.

In conversation with Indigenous knowledge and in the shadow of the plantation, Degener, Wiebke, Pohina, Iwasaki, Park, and dozens of other collaborators who collected for Degener's herbarium were literally uprooting Hawaiian plants, pulling them out of the ground and converting them into specimens with Latin names and according to Linnaean standards. Flattened, isolated in space, and frozen in time, these specimens were abstracted and alienated from their natural and cultural environments while acquiring economic value in exchanges.

Yet, a certain animacy—of the plant and its sociopolitical worlds—remains in these herbarium specimens. The plants' living qualities, communicated through the collectors' handwriting, sometimes jump right off of the page. In that same ʻeke silversword specimen from this essay's opening, Wiebke and Degener wrote that the plant was "creeping&branching&dying." Like Park's insistence that the yomogi was "escaping" a labor camp, the ʻeke silversword description demonstrates a careful observance of the plant's agency. The plant's agency reaches beyond the confines of

FIGURE 8.7

Otto Degener, Yasumi Iwasaki, and Oliver Pohina collected this rare specimen of *Bidens hillebrandiana* subsp. *hillebrandiana*, endemic to the Island of Hawai'i and now critically endangered. Perhaps denoting the specimen's value even in 1930, Degener listed, at the top of the label, the other collectors and herbaria to which he sent duplicates, including botanist Earl Edward Sherff, the New York Botanical Garden, Gray Herbarium at Harvard University, and the Wellesley College Herbarium. Bishop Museum Herbarium Pacificum, Honolulu.

the white page and temporal moment; it haunts both the colonial moment in which it was collected and the science archive that wants it static and dead.

Moreover, in documenting an increasingly rare endemic plant, Wiebke and Degener were writing the specimen into a narrative of the "vanishing native," or the apparent decline of Indigenous ways of life under U.S. settler colonialism more generally.[48] As the plant was "creeping& branching&dying," so, too, was Hawai'i and the Native Hawaiians of old, at least according to Degener's line of thinking. He sought to preserve endemic plants like the silversword while also trying to "save" Kanaka Maoli men like Wiebke. Degener and Wiebke began a gradual breakdown in 1928, when Wiebke dropped out of university and temporarily discontinued his medical studies. Degener disapproved of Wiebke's decision and advised him to continue his medical education, if not for himself, then for the good of "the Hawaiian race." Wiebke's reply suggests that he was tired of seeking the approval of white settlers, who, he knew, expected little from Kānaka Maoli. Instead, Wiebke was going to live a life according to his own measure of happiness. He wrote:

> You mentioned in your letter that I should sacrifice myself for the sake of the Hawaiian race. I am not going to do anything of that sort. Whatever success I may make will be for the sake of (me) and mine. The reputation of the Hawaiians are so far gone that there is no hope for them. If I made a success of myself, it will be credited to the German that's in me. If I make a failure of my life it will be because of the Hawaiian, so what's the use . . .
>
> I leave this martyr business to some other more worthy (or rather more stupid) than myself. Whatever happiness I can derive from this existence, I don't intend to miss.[49]

His reply suggests that he was keenly aware of how white settlers mapped success onto racial hierarchies and conceptualized Kānaka Maoli as a doomed people—just as Degener was beginning to conceptualize native Hawaiian plants as doomed species. When placed in the context of Wiebke and Degener's larger relationship, the ʻeke silversword specimen not only elicits Wiebke's erasure (his crossed-out name), but it also presents a colonial haunting, of the destructive acts and narratives that impacted Native lives, human and nonhuman. Similar to how Black feminist scholars like Tiffany Lethabo King have taken seriously "the unintended and ghostly reproduction" of indigo spots on eighteenth-century white settler maps, I argue that these specimen notations can be read "as a haunting and a reminder" of the force of Native life. As in King's maps, the "provisional, informal, or phantasmagoric details" of the specimen help us to, in Katherine McKittrick's words, "betray the archive," to read it "for what else happened" besides death.[50]

Conclusion: Reading the Herbarium as an Evolving Archive

Another haunting of the colonial conditions of both the historical collecting of these specimens and their contemporary presence in colonial institutions is the fact that these coveted herbarium specimens are in a state of slow decay. Those same silversword specimens that Degener, Wiebke, and their collaborators collected and filed away in the Bishop Museum in the 1920s and 1930s are, today, slowly crumbling. Evidence of the plants' ongoing material transformation and agency lies in the specimens' constant threat of breaking apart, changing color, and otherwise decaying from their original forms. As the U.S. occupation of the Hawaiian Islands and their ecological destruction are ongoing, Hawaiian herbarium specimens are just one "rot that remains" from U.S. settler colonialism, including the colonial practice of botany.[51]

This constant state of decay reminds us that finding elasticity in the herbarium also means recognizing that the archive itself is an evolving creature that plays an ongoing role in Indigenous erasure, colonial legacies, and political possibilities. Wiebke's crossed-out name on paper has translated into erasure in the virtual realm as herbarium specimens are digitized. In the last few decades, the mass digitization of specimens and their potential for global, open-access sharing has promised a means to "democratize" and "decolonize" botanical collections and remedy past colonial extractions of plants.[52] Yet, the digitization process can also risk reproducing historical erasures and incommensurabilities. This has been the case with Wiebke, but also with some women collectors like Mary Vaux Walcott. Hundreds of Walcott's specimens from the early twentieth century, which were labeled as collected by "Mrs. C. D. Walcott," were subsumed under her husband's name and collection ID during the digitization process at the National Museum of Natural History until a summer intern noticed the automated error.[53] While digitization and the ongoing use of these collections more generally present opportunities for interventions, the outcomes of those interventions are hardly predetermined to rectify past injustices.

This is why it is so important to encourage humanities scholars to utilize herbarium specimens, reinterpreting and refashioning their meanings, just as scientists have reimagined them as sources of DNA sampling and clues to address climate change. Opening up the science archive to humanists means finding new ways to read specimens according to past and present sociopolitical worlds; it also means stretching the archive to read for silences, signs of life, and hauntings beyond what is materially obvious and beyond existing narratives. Reading specimens for dynamic contexts and ongoing change in the herbaria means resisting the stench of death in the science archive and instead breathing life and possibilities into it.

Notes

1 Note that the plural of "Kanaka Maoli," meaning "true people," is "Kānaka Maoli," with the diacritical mark kahakō (macron).

2 Alexa Merrill, email to the author, January 25, 2018.

3 *Flora Hawaiiensis* was published in seven loose-leaf volumes, with the latter volumes coauthored by Isa Degener. Otto Degener, *Flora Hawaiiensis: The New Illustrated Flora of the Hawaiian Islands* (Honolulu: n.p., 1932–).

4 Isa Hansen and Otto Degener met in 1952 and wed in 1953, more than thirty years after Degener's collaborations with Wiebke. Brian M. Boom, "The Role of The New York Botanical Garden in Plant and Fungal Conservation," *Brittonia* 68 (2016): 307, https://doi.org/10.1007/s12228-016-9421-9; and Annette Lykknes, Donald L. Opitz, and Brigitte Van Tiggelen, eds., *For Better or Worse? Collaborative Couples in the Sciences* (Basel: Birkhäuser, 2012).

5 Michaela Schmull, email to the author, September 22, 2022.

6 "Archives and Manuscript Collections: Otto Degener Records (1920–1987)," *The New York Botanical Garden*, https://sciweb.nybg.org/science2/libr/finding_guide/degener3.asp.html; and Boom, "The Role of The New York Botanical Garden in Plant and Fungal Conservation," 307.

7 Deborah Woodcock, "To Restore Watersheds: Early Twentieth-Century Tree Planting in Hawai'i," *Annals of the Association of American Geographers* 93, no. 3 (2003): 632.

8 For examples of scholarship on "invisible labor," see Fiona Williamson, "Just Doing Their Job: The Hidden Meteorologists of Colonial Hong Kong c. 1883–1914," *The British Journal for the History of Science* 54 (2021): 341–59; Kapil Raj, "Go-Betweens, Travelers, and Cultural Translators," in *A Companion to the History of Science*, ed. Bernard Lightman (Oxford: John Wiley & Sons, 2016); Christopher M. Parsons, "The Natural History of Colonial Science: Joseph-François Lafitau's Discovery of Ginseng and Its Afterlives," *William and Mary Quarterly* 73, no. 1 (2016): 37–72; Benjamin R. Smith, "'Indigenous' and 'Scientific' Knowledge in Central Cape York Peninsula," in *Local Science vs. Global Science: Approaches to Indigenous Knowledge in International Development*, ed. Paul Sillitoe (New York: Berghahn Books, 2009); and Lydia Barnett, "Showing and Hiding: The Flickering Visibility of Earth Workers in the Archives of Earth Science," *History of Science* 58, no. 3 (2020): 245–74.

9 J'Nese Williams, "Plantation Botany: Slavery and the Infrastructure of Government Science in the St. Vincent Botanic Garden, 1765–1820," *Berichte zur Wissenschaftsgeschichte* 44, no. 2 (2021): 137–58; and James Delbourgo, *Collecting the World: Hans Sloane and the Origins of the British Museum* (Cambridge, Mass.: The Belknap Press of Harvard University Press, 2017).

10 Daniela Bleichmar, *Visible Empire: Botanical Expeditions and Visual Culture in the Hispanic Enlightenment* (Chicago: University of Chicago Press, 2012), 151.

11 This follows Bruno Latour's idea of immutable mobiles. Bruno Latour, "Visualisation and Cognition: Drawing Things Together," *Knowledge and Society: Studies in the Sociology of Culture and Present* 6 (1986): 7–9; and Bleichmar, *Visible Empire*, 151.

12 Marisa J. Fuentes, *Dispossessed Lives: Enslaved Women, Violence, and the Archive* (Philadelphia: University of Pennsylvania Press, 2016), 136. For more on reading colonial archives creatively, see Michel-Rolph Trouillot, *Silencing the Past: Power and the Production of History* (Boston: Beacon Press, 2015); Saidiya Hartman, "Venus in Two Acts," *Small Axe* 12, no. 2 (2008): 1–14; and Ann Laura Stoler, ed., *Imperial Debris: On Ruins and Ruination* (Durham: Duke University Press, 2013).

13 Kathleen Cruz Gutierrez, "*Cycas wadei* and Enduring White Space," in *Empire and Environment: Ecological Ruin in the Transpacific*, ed. Jeffrey Santa Ana, Heidi Amin-Hong, Rina Garcia Chua, et al. (Ann Arbor: University of Michigan Press, 2022), 76.

14 Gutierrez, "*Cycas wadei* and Enduring White Space," 74.

15 Adding chemicals to, or "poisoning," plant specimens was a common practice among contemporary botanical collectors to ward off pests and moisture. Otto Degener to N. L. Britton, Director of New York Botanical Garden, August 22, 1928, Series 2, Box 5, Folder "Britton, N. L.," RG4: Otto Degener Records (hereafter ODR), The Archives of The New York Botanical Garden, New York Botanical Garden, Bronx, N.Y. (hereafter NYBG).

16 Otto Degener to James J. White, December 26, 1981, Series 1, Box 2, Folder "Illustrators/houseboys," ODR, NYBG; and Otto Degener to unknown receiver, date unknown [ca. 1981?], Series 1, Box 2, Folder "Illustrators/houseboys," ODR, NYBG.

17 The Immigration Act of 1924, or the Johnson–Reed Act, established a quota system for immigration based on national origins and effectively banned immigration to the United States from Asian countries, except for the Phillippines, which was a U.S. territory. Mae M. Ngai, *Impossible Subjects: Illegal Aliens and the Making of Modern America* (Princeton: Princeton University Press, 2004); Erika Lee, *At America's Gates: Chinese Immigration During the Exclusion Era, 1882–1943* (Chapel Hill: University of North Carolina Press, 2003); and Sucheng Chan, *Asian Americans: An Interpretive History* (Boston: Twayne, 1991).

18 Otto Degener to James J. White, December 26, 1981, Series 1, Box 2, Folder "Illustrators/houseboys," ODR, NYBG; and "Park to Teach Oil Painting Classes at YMCA," *Honolulu Advertiser*, September 18, 1948.

19 Otto Degener to unknown receiver, date unknown [ca. 1981?], Series 1, Box 2, Folder "Illustrators/houseboys," ODR, NYBG; and Otto Degener to James J. White, December 26, 1981, Series 2, Box 9, NYBG.

20 Otto Degener, undated letter [ca. early 1980s], Series 1, Box 2, Folder "Illustrators/houseboys," ODR, NYBG; and "Notes, News, and Comment," *Journal of the New York Botanical Garden* 34 (December 1933): 276.

21 Dawn Bohulano Mabalon, *Little Manila Is in the Heart: The Making of the Filipina/o American Community in Stockton, California* (Durham: Duke University Press, 2013), 40–49.

22 Otto Degener to James J. White, December 16, 1981, Series 1, Box 2, Folder "Illustrators/houseboys," ODR, NYBG; and Otto Degener, undated letter [ca. early 1980s], Series 1, Box 2, Folder "Illustrators/houseboys," ODR, NYBG.

23 Otto Degener, undated letter [ca. early 1980s], Series 1, Box 2, Folder "Illustrators/houseboys," ODR, NYBG.

24 Gary Y. Okihiro, *Cane Fires: The Anti-Japanese Movement in Hawaii, 1865–1945* (Philadelphia: Temple University Press, 1999), 138–61, 187.

25 Common names for *Artemisia princeps*, native to China, Japan, and Korea, include yomogi, Japanese or Korean mugwort, or Korean wormwood. BISH, O'ahu: Kawela Bay, O. Degener 9694 (BISH [73455]).

26 Gary Y. Okihiro, *Pineapple Culture: A History of the Tropical and Temperate Zones* (Berkeley: University of California Press, 2009), 2–4.

27 Okihiro, *Pineapple Culture*, 110–29, 153–63.

28 BISH, Kauai: West slope of Omoe, near Koloa, O. Degener 27445 (BISH [75090]); and BISH, O'ahu: Near Kipapa Gulch, O. Degener 27859 (BISH [77658]).

29 For more on the plot in the context of the plantation, see Sylvia Wynter, "Novel and History, Plot and Plantation," *Savacou*, no. 5 (1971): 95–102; Katherine McKittrick, "Plantation Futures," *Small Axe* 17, no. 3 (2013): 1–15; Williams, "Plantation Botany"; and Judith A. Carney, "Botanical Gardens of the Dispossessed," in *In the Shadow of Slavery: Africa's Botanical Legacy in the Atlantic World*, by Judith A. Carney and Richard N. Rosomoff (Berkeley: University of California Press, 2009), 123–38.

30 BISH, O'ahu: Kawela Bay, O. Degener 9593 (BISH [57363]); and BISH, O'ahu: Kailua, O. Degener 5995 (BISH [47320]).

31 Otto Degener to James J. White, December 16, 1981, Series 1, Box 2, Folder "Illustrators/houseboys," ODR, NYBG.

32 Ronald T. Takaki, *Pau Hana: Plantation Life and Labor in Hawaii, 1835–1920* (Honolulu: University of Hawai'i Press, 1983), 60–66.

33 Nayan Shah, *Contagious Divides: Epidemics and Race in San Francisco's Chinatown* (Berkeley: University of California Press, 2001), 89; and Henry Yoshitaka Kiyama, *The Four Immigrants Manga: A Japanese Experience in San Francisco, 1904–1924*, trans. Frederik L. Schodt (Berkeley, Calif.: Stone Bridge Press, 1999), 9.

34 BISH, O'ahu: Kawela Bay, O. Degener 9693 (BISH [73455]).

35 Lilikalā Kame'eleihiwa, *Native Land and Foreign Desires: Pahea Lā E Pono Ai? How Shall We Live in Harmony?* (Honolulu: Bishop Museum Press, 1992), 2. For more on the significance of genealogy to Native Hawaiian cosmology and identity, see J. Kēhaulani Kauanui, *Hawaiian Blood: Colonialism and the Politics of Sovereignty and Indigeneity* (Durham: Duke University Press, 2008); Noenoe K. Silva, *Aloha Betrayed: Native Hawaiian Resistance to American Colonialism* (Durham: Duke University Press, 2004); and Lilikalā Kame'eleihiwa, "Kalana O'ahu me he 'Āina Momona," in *Food and Power in Hawai'i: Visions of Food Democracy*, ed. Aya Hirata Kimura and Krisnawati Suryanata (Honolulu: University of Hawai'i Press, 2016), 59.

36 BISH, O'ahu: Northeast of Nuuanu Pali, O. Degener 5285 (BISH [74226]).

37 BISH, O'ahu: Makapu'u Point near abandoned native city on way to Waimanalo, O. Degener 6072 (BISH [47830]).

38 BISH, O'ahu: Wailau Valley, just mauka of native shacks, O. Degener 8950 (BISH [117641]).

39 For more on Kanaka Maoli geographies and spatial orientations, see David A. Chang, *The World and All the Things Upon It: Native Hawaiian Geographies of Exploration* (Minneapolis: Minnesota University Press, 2016); and Carlos Andrade, "A Hawaiian Geography or a Geography of Hawai'i?," in *I Ulu I Ka 'Āina: Land*, ed. Jonathan Kay Kamakawiwo'ole Osorio (Honolulu: University of Hawai'i Press, 2014), 4–22.

40 Kanaka Maoli botanist Isabella Aiona Abbott similarly describes how 'ama'u fronds were used in house thatching in her classic ethnobotanical text. Note that for this word and others, I added the 'okina (diacritical mark indicating glottal stop) to Wiebke's "amau" to reflect the contemporary spelling, 'ama'u. Isabella Aiona Abbott, *Lā'au Hawai'i: Traditional Hawaiian Uses of Plants* (1992; reprint, Honolulu: Bishop Museum Press, 2019).

41 Henry Wiebke to Otto Degener, n.d. [ca. 1928–1931], Series 2, Box 9, Folder "Wiebke," ODR, NYBG.

42 Pohina was born in 1906 in Hawai'i. His grandfather was also a fisherman in Kona. United States Census Bureau, *Fifteenth Census of the United States, 1930* (Washington, D.C.: National Archives and Records Administration, 1932); *1930 United States Federal Census* [database online]; and United States Census Bureau, *Fourteenth Census of the United States, 1920* (Washington, D.C.: National Archives and Records Administration, 1923); *1920 United States Federal Census* [database online].

43 Oliver Pohina to Otto Degener, n.d. [ca. October 1929], Series 2, Box 9, Folder "Pohina," Degener (RG4), Archives, NYBG.

44 Oliver Pohina to Otto Degener, December 1929, Series 2, Box 9, Folder "Pohina," Degener (RG4), Archives, NYBG.

45 Oliver Pohina to Otto Degener, n.d. [ca. October 1929], Series 2, Box 9, Folder "Pohina," Degener (RG4), Archives, NYBG.

46 Oliver Pohina to Otto Degener, n.d. [ca. February 1929], Series 2, Box 9, Folder "Pohina," Degener (RG4), Archives, NYBG.

47 BISH, Hawai'i [Island]: Along ditch trail in Pololū Valley, O. Degener 3090 (BISH [74176]); and BISH, Hawai'i [Island]: Kohala District: Niulii, O. Degener 3901 (BISH [13844]).

48 While this trope of the "vanishing native" was popular at the time, it ignores the fact that Indigenous peoples, including Kānaka Maoli, maintained important continuities in population, culture, and knowledge and resisted colonialism.

49 Henry Wiebke to Otto Degener, October 1, 1928, Series 2, Box 9, Folder "Wiebke," ODR, NYBG.

50 Tiffany Lethabo King, *The Black Shoals: Offshore Formations of Black and Native Studies* (Durham: Duke University Press, 2019), 78–79; and Katherine McKittrick, "Diachronic Loops/Deadweight Tonnage/Bad Made Measure," *cultural geographies* 23, no. 1 (2015): 3.

51 I take this phrase from poet Derek Walcott, quoted in Stoler, *Imperial Debris*, 1–2.

52 Daniel S. Park, Xiao Feng, Shinobu Akiyama, et al., "The Colonial Legacy of Herbaria," *bioRxiv* (2021), https://doi.org/10.1101/2021.10.27.466174.

53 Thank you to Dr. Maura Flannery for bringing this case to my attention. Rebecca Dikow and Megan Glenn, "What's in a Name? Uncovering Stories of Women in Science," *Smithsonian*, April 14, 2020, https://datascience.si.edu/news/whatsinaname.

9

DISRUPTIVE TAXONOMIES

In the Herbarium with Temitayo Ogunbiyi

YOTA BATSAKI

Natural history collections are organized assemblages of plant and animal specimens, minerals, and cultural artifacts that double as archives of biodiversity and human diversity. Historically determined and, therefore, open to contestation and revaluation, they disclose how practices of inventorying and representing the natural world both inform and reflect broader epistemic and political shifts.[1] Among such collections, the herbarium is perhaps one of the oldest and least well-known outside the circles of botanical and taxonomic expertise.[2] Defined as "a documented collection of dried plants permanently preserved on paper," the herbarium has its roots in the sixteenth century, when it served as a memory aid and pedagogical tool for the emerging botanist, often in tandem with the botanical garden where the living counterparts of the dried specimens grew.[3] From their roots in observation and teaching, herbarium specimens morphed into records for scientific identification and classification. But the rise of colonial botany tilted herbaria toward the instrumentalization of Western scientific practices at the service of maximizing profit. Herbaria often reside in centers of colonial science and knowledge accumulation or were fed by the networks that formed around these centers of imperial control. Moreover, their holdings often memorialize the name of the agent of empire who brought the specimens to the metropolitan center while obscuring the local voices and knowledges that enabled the "discoveries" of rare or useful plants. The concentration of major herbaria in capitals such as London, Paris, and Washington, D.C., far removed from the biodiversity that they record, underscores their colonial legacies and their associations with massive projects of relocating the world's floras for the benefit of economic botany and bioprospecting.[4] The herbarium's organizational principles and epistemic goals can thus appear hegemonic and extractive, cultivating a misleading appearance of objectivity and control over the natural world. Indeed, the herbarium's techniques of preservation and transmission have been presented by historians of science and empire as paradigmatic of the practices that, in the modern period, were employed to inventory nature as a preamble to its ruthless exploitation.

In *The Order of Things*, Michel Foucault analyzes how living beings were transformed into specimens through the practices of modern taxonomy. He situates these practices in the eighteenth century, "the Age of Classifying," embodied in the figure of Carl Linnaeus and expressed in his binomial system. Linnaeus established the hierarchy of class, order, genus, and species and introduced the convention of a two-word Latin name—designating the genus and species—that remains the standard. Linnaeus's taxonomic revolution was dependent on global projects of collecting, as he instructed his disciples to send him plants from Europe, Asia, Africa, and the Americas. Scientific naming was predicated on prior extraction: the removal of organisms from local relationships, affinities, or narratives for their preservation and insertion, as things, in the drawers of systematic taxonomy. Foucault is less focused on the undoing of ecological relationships and more interested in how the new taxonomy wrested creatures from the web of discourse. Organisms were no longer understood through the stories in which they were enmeshed but rather according to their position in the taxonomic scheme and its spatial instantiations: "The documents of this new history are not other words, texts or records, but unencumbered spaces in which things are juxtaposed: herbariums, collections, gardens . . . a non-temporal rectangle in which, stripped of all commentary, of all enveloping language, creatures present themselves one beside another, their surfaces visible . . . and bearers of nothing but their own individual names."[5] Achieving this surface visibility required a "limiting and filtering" of elements deemed worthy of observation and recording: the willingness to "be content with seeing—with seeing a few things systematically."[6] The "new field of visibility" of natural history, therefore, entailed, for Foucault, a scraping away of opacity and complexity.[7] As exemplified by the herbarium specimen, the plant lent itself to this sorting and flattening effect more easily than the animal: "Hence the epistemological precedence enjoyed by botany."[8] Foucault intimates that this filtering and flattening process is violent and intrusive, a form of representation in which "the living being . . . appears as though stripped naked."[9] In the case of the herbarium sheet, the plant is prized away from soil and multispecies relationships as well as local names and uses, and often gives up its color, fragrance, and habit of growth. What is left is the minimal structure required for the purposes of taxonomy—and that structure is then carefully, even lovingly, preserved, ideally in perpetuity.

As Linnaeus's system of binomial classification attained dominance, plants became paradigmatic of the new discursive order. The annotation on the herbarium sheet preserves the bare minimum that is needed to record and stabilize the specimen, then render it mobile and exchangeable: species name, date and site of collecting, and collector's name. The herbarium's economy of means thus becomes a stand-in for the entirety of botanical collecting; its allure haunts and shapes Foucault's account of epistemic rupture, even though the products of natural history expeditions would also include seeds, field notes, sketches, travel diaries, and other materials. Indeed, as Staffan Müller-Wille and Isabelle Charmantier have shown, the data overload that accompanied the influx of specimens from the colonial enterprise was a generative force behind the information pruning that enabled the formation of vast archives, such as herbaria.[10] The desiccation and portability of the herbarium specimen also fit in well with Bruno Latour's account of how practices of stabilization and mobilization fed the circulation of knowledge between "centers of calculation" and colonial spaces of collecting.[11] And Lorraine Daston has identified in the convention of the herbarium "holotype"—the single specimen that the original giver of the scientific name used as the type for the description of the species—the core of botany's particular and successful "art of transmission." In Daston's account, the type specimen is the handy personification of the plant: "the face—the desiccated, flattened face to be sure, but still the face—that is attached to the name of a species, and on the permanence of that relationship depends the transmission of botanical knowledge amassed for centuries."[12] The hyperbole and personification in Daston's description attest to the attraction, for the historian, of the dried specimen as a surrogate for modern botanical knowledge practices.

Yet Foucault's formative account is not borne out by the experience of the herbarium. Elements of discourse survive in handwritten or typed comments on the sheet, occasionally recording not only the location, date, and collector but also successive renamings and reorganizings, the rise and fall of taxonomic systems, figures in the margins of professional or colonial botany, and, more recently, local names, uses, and ecological contexts. The organizational purity suggested by "unencumbered spaces" is belied by the perennial problem of storage and the irrepressible intrusion of hybridity: sketches and botanical illustrations or print and manuscript material may be interspersed with the dried specimens or filed nearby. The threat of disorder lurks in piles of folders and boxes of uncatalogued materials in the backlog, puncturing the illusion of mastery. Nor is the herbarium simply the clinical space of abstract ratiocination evoked by the "non-temporal rectangle" of the taxonomic table. Rather, the archive of desiccated plants has powerful sensory, affective, and aesthetic dimensions that envelop and trouble the visitor. The residual scent of dried plants and paper mingles with the tactile pleasure of the sheet, the rustle of paper, and the mix of desire and fear in handling a fragile specimen. The pathos of crumbling parts and fading colors is heightened by the passing of time inscribed in the date: sometimes an intimation of organisms and perhaps entire ecosystems that have vanished. Ecological grief may well up in the encounter with the remnants of a species that has become extinct in the wild or extirpated from its local habitat.[13] The lineaments of the herbarium specimen thus have the capacity to evoke, in miniature, the poetics of the ruin.

Arguably, the aesthetic of the herbarium goes beyond the sensory qualities of the desiccated plant and its paper container: it also inheres in the archive's particular mode of presentation. This aesthetic is evident in the herbarium sheet's economy of means, which seeks to capture the taxonomic essence of the organism by pressing, bending, folding, and reversing—in its practiced filtering of the visible. The multimedia nature of the sheet, which often includes various specimen parts, label(s), and annotation(s), partakes of the method of collage, an assemblage that strives for synthesis while divulging foundational dismemberments and sutures. The supplement may make its appearance as a little envelope attached to the sheet that holds seeds or detached parts. The different methods of attaching the plant, which may include glue, transparent resin, little bands of paper, and even thread, contribute to this pasted and sutured effect (sometimes slapdash, often meticulous) and its precarious hold on the plant. Lastly, the necessity of the part standing for the whole—a twig, say, subbing for the majestic tree—speaks to the herbarium's metonymic function and power: it activates the imagination to restore aliveness to the dead and wholeness to the fragment.

Because the herbarium is a site of both preservation and loss, combining practices of extraction and care, it has attracted several contemporary artists who are interested in the legacies of natural history at a time of environmental devastation. Desiccated ferns framed in the manner of herbarium specimens encircle coal-spewing bunkers in Anselm Kiefer's *Secret of the Ferns* (2007), a monumental installation in the Margulies Collection in Miami that eerily evokes the slow violence of extinction.[14] While Kiefer's work gives extinction a planetary feel, other artists tap into personal herbaria to trace and commemorate the local extirpation of species due to human-driven change. An example is Leah Sobsey's and Robin Vuchnich's *In Search of Thoreau's Flowers: An Exploration of Change and Loss*, an exhibition that opened in 2022 at the Harvard Museum of Natural History and was inspired by Henry David Thoreau's collection of pressed plants. As a paradigmatic site for the organization of knowledge about the natural world according to principles of Western science, the herbarium has also attracted artists interested in the history of natural history. Mark Dion's *Herbarium Perrine* (1996–) is the fictional reconstruction of the herbarium of Henry E. Perrine, a nineteenth-century naturalist and proponent of tropical crop introductions to Florida who perished with his collections in the conflict with the Seminoles. In this case, the artwork fills the gap of a vanished archive:

a placeholder for past violence, dispossession, and botanical ambition. In these aesthetic ventures, the herbarium becomes a contested site for exploring emotions of curiosity and wonder and modes of care and preservation, alongside exploitative practices of colonialism and capitalism that view nature through the lens of resource extraction.

The immersive experience of the herbarium informs the practice of Temitayo Ogunbiyi, a contemporary artist and curator based in Lagos, Nigeria. Ogunbiyi is interested in collecting as a vehicle of colonialism, and her work often engages with plants as pathways into histories of migration and belonging. She sometimes adopts the mode of presentation of the natural history collection to critique and reframe it. In a 2023 exhibition at the Museum Tinguely, Basel, entitled *You will find new framing in crafts of old*, she populated the gallery with furniture that resembled wood and glass cabinets from a natural history museum but that was disconcertingly scattered around the room or suspended from the walls (Figure 9.1) The drawers were filled with artworks rather than specimens, their inspiration derived from the artist's personal experience with plant collections. As a Smithsonian artist-in-residence in 2018, Ogunbiyi spent time in the herbarium of the National Museum of Natural History in Washington, D.C. This experience inspired the artist to interrogate the conventions of the herbarium both as an archival system and as a mode of representation. Although Ogunbiyi originally planned to research specific plant histories, during the residency she became interested in the architectural structure and organization of the herbarium as a site. Specifically, she noted how the set dimensions of the herbarium sheet (approximately 42 × 29 cm or 16.5 × 11.4 in) dictate, in turn, the sizes of the folders, drawers, cabinets, and rooms.[15] Both content and container reflect a desire to standardize and render measurable, mobile, and exchangeable the specimens in ways that enable control of the collection and ultimately the world.

Spatial organization as the concrete embodiment of institutional practices of surveillance and control brings us back to Foucault. If the herbarium sheet is the cornerstone of the herbarium's organization, its materiality dictates the mode of presentation, resulting in the flattened and decontextualized specimen. If life exists

FIGURE 9.1
Temitayo Ogunbiyi, *You will find new framing in crafts of old*. Installation view, Museum Tinguely, Basel, 2023. Photograph by Bettina Matthiessen.

FIGURE 9.2
Temitayo Ogunbiyi, *You will find new framing in crafts of old*, detail, 2023. Pencil and watercolor on herbarium paper, 16.5 × 11.5 in. Commissioned by the Museum Tinguely. Photograph by Bettina Matthiessen.

in three dimensions, the sheet exemplifies its two-dimensional deadening not merely as a preservation technique but as an epistemic modality. The rectangle of set proportions thus maps the disciplinary and ideological confines of a certain kind of scientific gaze (extrapolating, exemplifying, disconnected from biosystems, and historically complicit in the imperial project) that Ogunbiyi's practice sought to interrogate. Yet, during her Smithsonian residency, Ogunbiyi "fell for herbarium paper."[16] Her experience in the United States National Herbarium inspired a series of drawings and paintings on herbarium paper that complied with the material's texture and weight—qualities to which the artist initially had to adjust her technique—as well as its dimensions. The artist's choice of pencil and watercolor or acrylic to represent the plant on the sheet, instead of affixing its organic remains, is a form of reanimation. Similarly, the substitution of enigmatic phrases in cursive (which double as the titles of the works) for the Latin binomial reintroduces an element of storytelling. The substrate of the deadening and flattening operations of the herbarium thus becomes a springboard for botanical reimaginings, transmuting archival constraint into an engine of creativity.

Many of Ogunbiyi's subsequent paintings of plants abide by a tradition of botanical representation that encompasses both the herbarium sheet and the botanical watercolor or scientific illustration and that presents the plant against a blank background. Yet, at the same time, these paintings on herbarium paper undo the fantasy of visibility and the imperative to serve plant identification that underpin both the flattened specimen and the two-dimensional illustration. Some works, such as *You will find new framing in crafts of old* (2023), trouble identification by blurring the morphology of the plant, restoring to the organism its living fluidity (Figure 9.2). The painting's

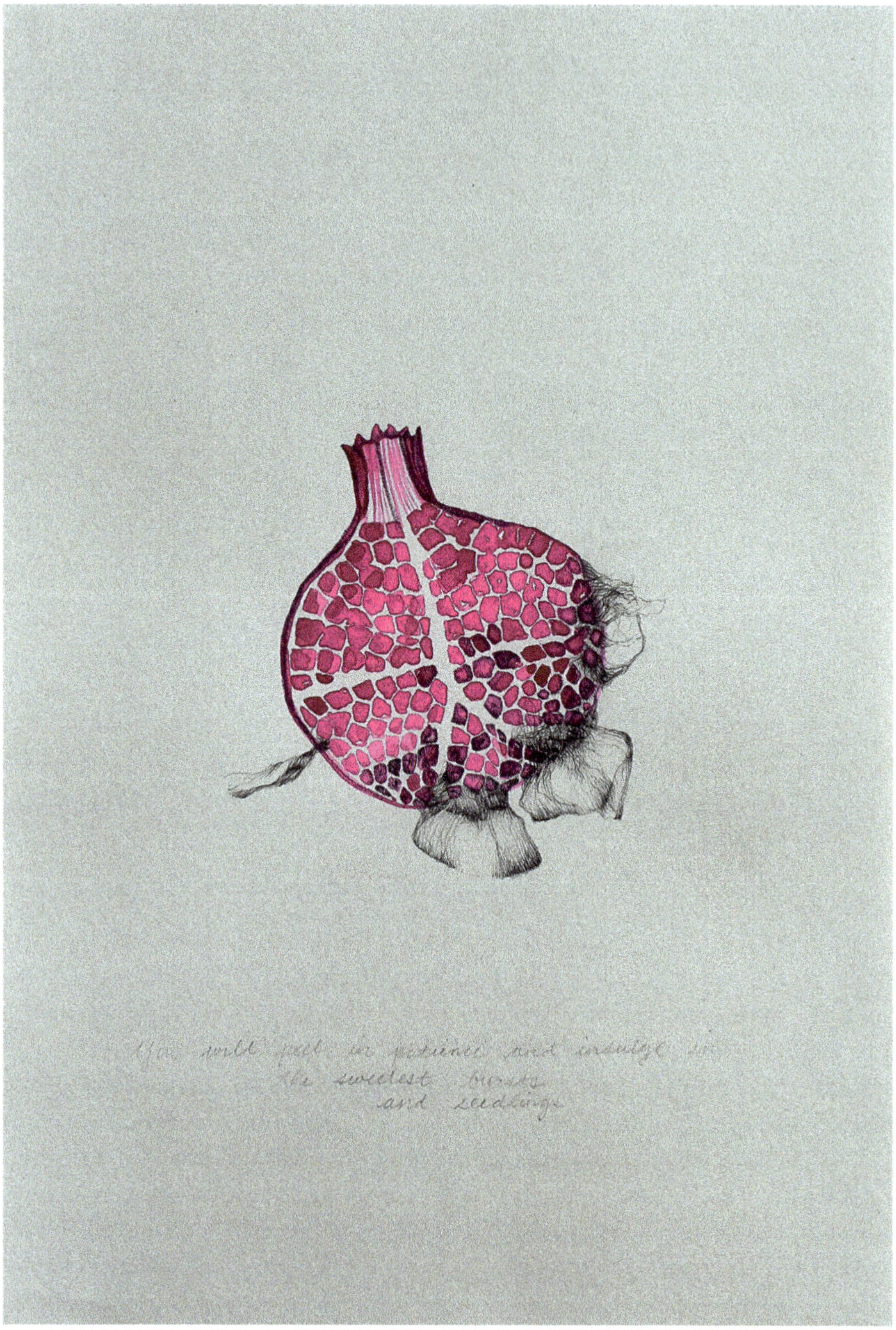

FIGURE 9.3
Temitayo Ogunbiyi, *You will peel in patience and indulge in the sweet bursts and seedlings*, 2022. Pencil and acrylic on herbarium paper, 16.5 × 11.5 in. Photograph by Bettina Matthiessen.

indeterminacy—the plant is kohlrabi radish—disrupts the tradition's aspiration to metonymic fullness of representation in the service of mastery. Instead, Ogunbiyi's works on herbarium paper restore to the plant its color, character, and even its generative potential. *You will peel in patience and indulge in the sweet bursts and seedlings* (2022) painstakingly and invitingly depicts the seeds of the pomegranate. The title evokes the labor needed to get at the sweet reward, perhaps also alluding to the six-month wait that the mythical Persephone must endure in the underworld before the coming of spring and rebirth (Figure 9.3). Contrary to the standard instructions given to eighteenth- and nineteenth-century collectors to bring back examples of plant types in their integrity, the artist occasionally draws seemingly damaged "specimens." *You will always remember how you feel no matter the*

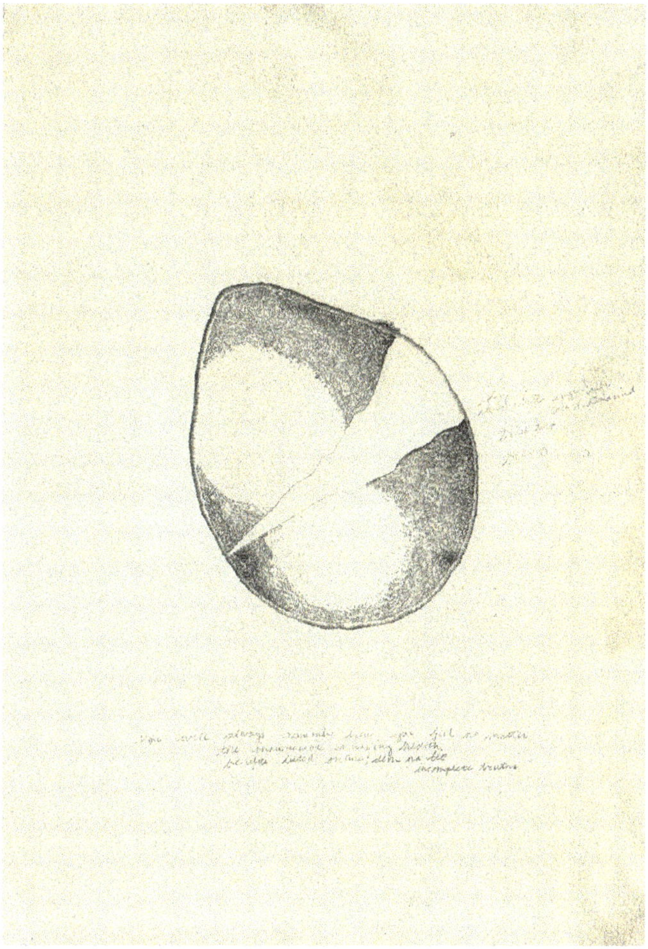

FIGURE 9.4
Temitayo Ogunbiyi, *You will always remember how you feel no matter the innumerable benefits listed online—dem na incomplete truths*, 2019. Pencil on herbarium paper, 16.5 × 11.5 in.

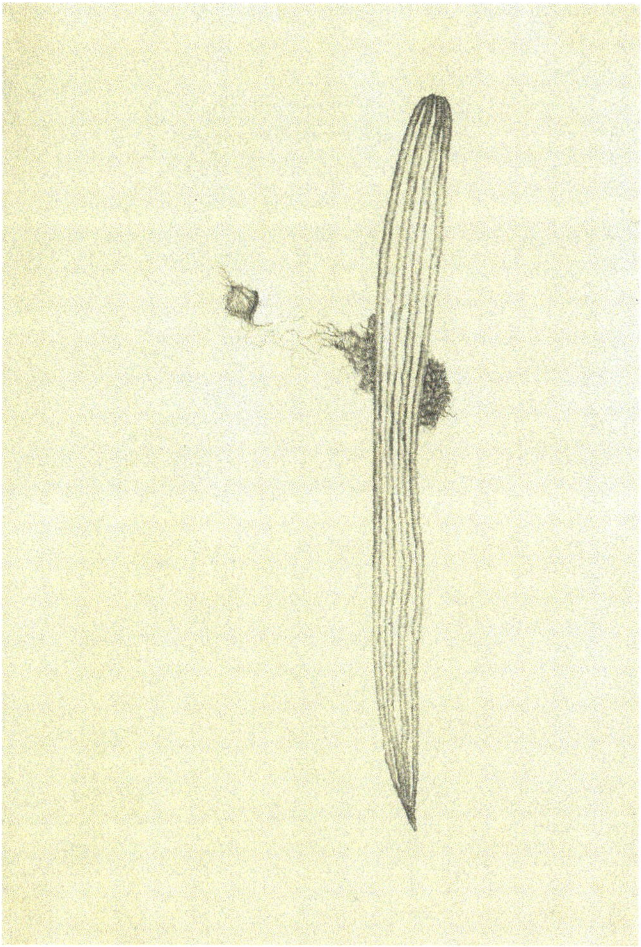

FIGURE 9.5
Temitayo Ogunbiyi, *You will chew the bittersweet knowing there might be some benefit*, 2019. Pencil on herbarium paper, 16.5 × 11.5 in.

innumerable benefits listed online—dem na incomplete truths (2019) goes against the often-idealized aesthetic of botanical representation to acknowledge the living organism's exposure to the world (Figure 9.4). The dismembering of the plant to maximize its visibility in the limited real estate of the herbarium sheet is lovingly countered in *You will chew the bittersweet knowing there might be some benefit* (2019) (Figure 9.5). Here, the seed of the moringa tree leaves its pod like a miniature spacecraft on its way to adventure, while delicate filaments maintain the traces of its relationship to the mother ship. This exquisite drawing in pencil throws into relief the uncanny fixity of the dried specimen, and perhaps by extension the fallacy of conceiving plants as sessile organisms. The artwork restores to the plant its potentialities of connectivity, generation, travel, and growth.

One of the dark legacies of the taxonomic impulse was the translation of the classificatory impulse into a hierarchy of beings that ultimately served racial ideologies and colonial violence.[17] A series of drawings by Ogunbiyi subtly refers to this taxonomic violence, only to undo it by blurring the distinction between the human and nonhuman. In the work of contemporary artists, many from the African diaspora, plant-human hybrids of the Anthropocene call on us

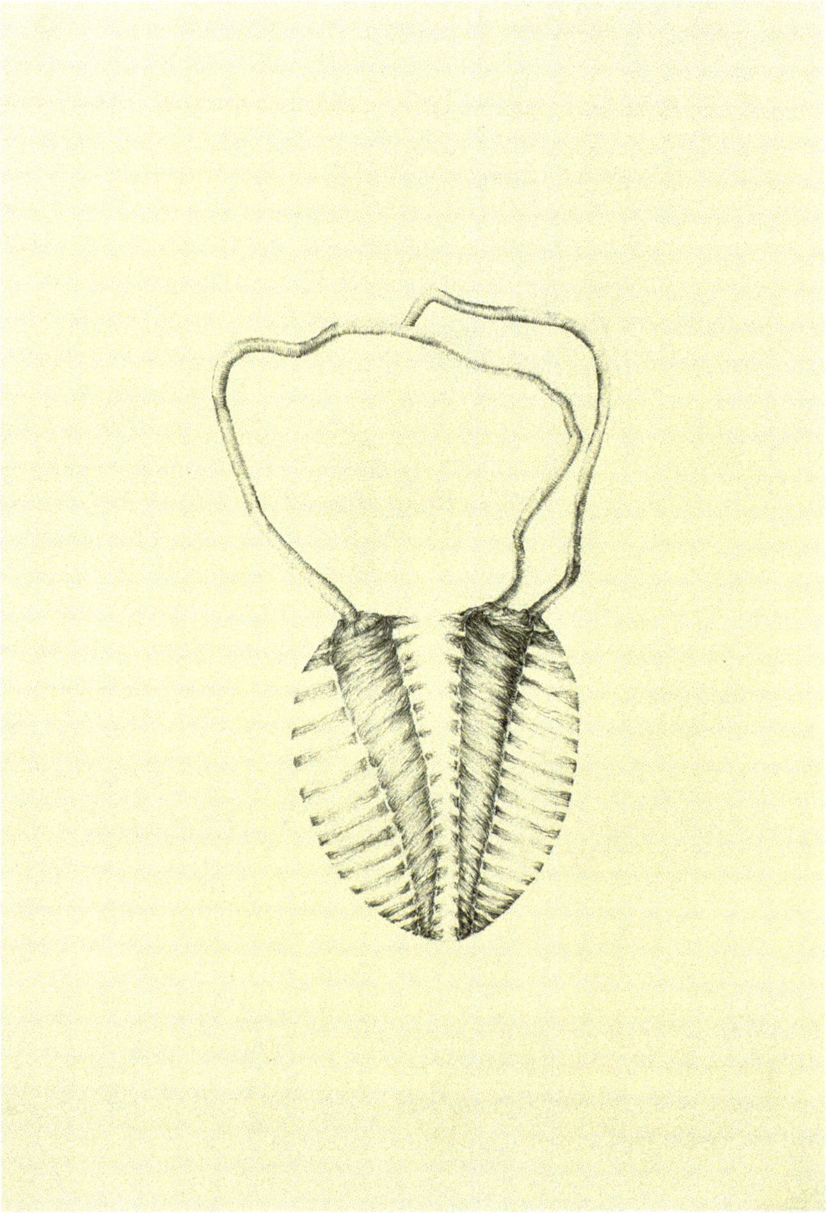

FIGURE 9.6
Temitayo Ogunbiyi, *You will know balance beyond the physical*, 2020. Pencil on herbarium paper, 16.5 × 11.5 in.

to reimagine our embeddedness in the natural world while alluding to the violent histories that brought us here.[18] Ogunbiyi's drawing, *You will know balance beyond the physical* (2020), accomplishes such a plant-human fusion through the depiction of hairstyles inspired by botanical patterns (Figure 9.6). The work evokes the artistic tradition of capturing Nigerian women's hairstyles, as in the images of J. D. 'Okhai Ojeikere, an acclaimed photographer who documented the artistic legacies of the country during the second half of the twentieth century. Plant-human analogies occur elsewhere in Ogunbiyi's work, as in the title of her 2020 solo exhibition, *Capillarité*, at 31Project, Paris, which referred to the vessels that circulate water in plants and blood in humans while also playing on the Latin root of capillary (meaning "hair" or "hairlike") and its affinity with the artist's delicate pencil strokes. That show also featured botanical patterns that are popular as traditional Nigerian hairstyles (Figure 9.7). The drawings were shown under glass, in the manner of rare and fragile natural history specimens. Whereas racial ideologies deployed the idea of a continuum with nonhuman

FIGURE 9.7
Temitayo Ogunbiyi, *Capillarité*. Installation view, solo exhibition at 31Project, Paris, 2020.
Photograph by Paul Nicoué.

beings as a strategy for the abjection and domination of humans, the artist taps into powerful ethical and aesthetic currents in contemporary thought that value interdependence and multispecies entanglements as a strategy of reparation. The scale of Ogunbiyi's drawings offers an alternative mode of registering the intricacy and fragility of plant-human entanglements and of relating to the nonhuman. *You will know balance beyond the physical* draws on traditions of both natural history illustration and ethnographic documentation; the indeterminacy of the drawing is an example of what Giovanni Aloi has called "the troubling ontological instability plants ignite."[19] The recursive image resists the hierarchical presentation of human versus plant, offering instead a portrait of symbiosis held in subtle and delicate balance.

Another means of distilling long histories of plant-human relationships and mobilities onto the herbarium sheet is through the association of plants with food and community. A painting of okra invites close looking at the plant's structure, while its title plays on the meaning of *diaspora* as a scattering of seeds. The annotation may also hint at the plant's companionship with violently displaced yet resilient African communities: *You will recall times when you can only imagine where the seeds have gone and how they've grown* (2020)

FIGURE 9.8

Temitayo Ogunbiyi, *You will recall times when you can only imagine where the seeds have gone and how they've grown*, 2020. Pencil and watercolor on herbarium paper, 16.5 × 11.5 in. Photograph by Raisa Galofre.

(Figure 9.8).[20] Okra (from the Igbo "okuru") is an African food staple that doubles as a symbol of African diasporic survival and the continuity of food traditions across the Atlantic.[21] Another painting of the plant, *You will order taxonomies according to your days* (2021), weaves it into a hairstyle that interprets okra as it appears in an Igbo dish, okra nsala. The work maps intimate itineraries into a spiral and recalls oral histories and legends of the transmission of seeds from the native land across the Middle Passage by braiding them in hair (Figure 9.9).[22]

The penciled annotations in cursive are an important element of Ogunbiyi's herbarium-inspired works. They repurpose a feature of the early herbarium sheet, the handwritten plant name that was later replaced by the typed label. The artist's use of second-person address transforms the seemingly objective bestowal of a scientific designation into a mode of address tied to subjectivity. If the application of the Linnaean binomial took the place of the language that had been "deposited upon things by time," narrative returns in the cursive annotations on Ogunbiyi's

FIGURE 9.9
Temitayo Ogunbiyi, *You will order taxonomies according to your days*, 2021. Pencil and watercolor on herbarium paper, 16.5 × 11.5 in. Photograph by Raisa Galofre.

herbarium drawings.[23] The penciled phrases sound like an injunction; but the seeming imperatives are—in Nigerian discourse—a form of prayer or encouragement. They articulate a wish for the Other, expressed as a second-person address that externalizes the Other's wishes, desires, or fears. For example, "You will succeed in this endeavor" is a supportive and auspicious encouragement of a friend who has just expressed uncertainty and/or desire for that very endeavor. Foucault argues that "natural history can be a well-constructed language only if the amount of play in it is enclosed."[24] Ogunbiyi's artistic practice introduces play back into the construction of the herbarium sheet and, by extension, the classification system, through her suggestive, sometimes cryptic notes. The annotations bring back the rush of human subjectivity with its wishes, desires, memories, its partial knowledge, the senses ("You will chew the bittersweet knowing") and traditions ("your great grandmother's taboos").

Ogunbiyi's work *You will know and confront your great grandmother's taboos* (2019) is made in pencil on herbarium paper of standard

FIGURE 9.10
Temitayo Ogunbiyi, *You will know and confront your great grandmother's taboos,* 2019. Pencil on herbarium paper, 16.5 × 11.5 in.

dimensions (Figure 9.10). It refers to the prohibition on eating taro in her grandmother's Igbo village—even though taro or cocoyam is a food staple in Nigeria, the biggest producer of the crop. Ogunbiyi has described this work as "somewhat [of a] self-portrait for me because the botanical reference is taro or cocoyam, as some call it here in Nigeria . . . I was told my grandmother never ate it, that it was taboo in Aboh, her village in Delta, a state in East Nigeria . . . My father mentioned to me that perhaps someone long ago ate it and fell ill so it then was decided that it was not to be eaten. It is fascinating how one isolated incident can inform a whole belief system for generations."[25] The work conveys Ogunbiyi's interest in the historical contingencies and weak rationalizations that underpin belief systems. Even Indigenous taxonomies—such as the distinction between edible and inedible or poisonous plants—are not exempt from her critique of classification systems, their arbitrary demarcations, and necessary exclusions. While her artistic practice interrogates the herbarium as a site of colonialism and takes a critical view of Western systems of classification, her critique can also encompass local knowledges and taxonomies, such as food taboos that may reveal a community's inherent aversion to change or experimentation. As shown by Anatole Tchikine in this volume, these Indigenous and vernacular taxonomies are rarely erased but rather more often overlaid with scientific classifications, and they survive in the mixture of conservatism and innovation characteristic of many food traditions.

What can the artist's foray into the herbarium tell us about the present and future of these plant archives? Under the foundational influence of Foucault, and in the wake of urgent and important recent explorations of the archive as a colonial institution, plant collections may appear fixed in the historical conditions of their production. But the artist's "archive fever" attests to the fact that the materiality of the archive both depends upon and exceeds its historical determination. I adopt Jacques Derrida's concept of "archive fever" as a counterpart to Foucault's discussion of the archive as the site of epistemic rupture. Derrida troubles the *arche* (both "origin" and "authority") of the archive, recasting it as a project that combines aggression with care, preservation with destruction—an unstable container with a past, present, and open future.

Derrida does not take issue with Foucault's description of the archive as a spatial configuration that inscribes a system of power relations: the powers of gathering, classifying, assigning residency, and exercising control. The archive, he acknowledges, needs a "guardian" and a "localization," a "place" and a "law."[26] The archive also needs a technique of *repetition*, demonstrated in the consistent dimensions of the herbarium sheet; and a technique of *impression* and *inscription*, usually exemplified by the document or photograph, but here expressed as the pressing and annotating of the plant on the herbarium paper. Documents become an archive by means of this privileged topology that allows political power the control of memory. Drawing on psychoanalysis, Derrida acknowledges that this repetition compulsion is an instantiation of the death drive: of extraction, aggression, and destruction, coming into being "by originary finitude and expropriation."[27] *Archive Fever*, therefore, points to the foundational violence of the archive, as well as its capacity to lend itself to the economistic process of "the accumulation and capitalization of memory," the ongoing production of value from that foundational violence.[28] No wonder that historians of colonial botany have seen in the Western herbarium a "global mortuary."[29]

Yet Ogunbiyi's reclaiming of the herbarium sheet, and the powerful attraction that natural history collections exert on many contemporary artists more broadly, suggest the generative potential of the archive. Plant collections occupy a continuum of liveliness: from botanical gardens that display living plants, to seed banks that hold seeds in suspended animation and periodic resurrection, to herbaria that are filled with specimens presumed dead. The boundaries between them, however, are porous: herbaria may hold genetic material that contains the promise of reanimation, or botanical art that captures the likeness of the living plant alongside its skeletal remains, or a cryptic annotation that points to a historical figure previously marginal or unknown. For scientists interested in modeling climate change or shifting habitats and populations, herbaria are undergoing a revolution as treasure troves of biodiversity discovery.[30] These mixings and hauntings bring into question the archive's homogeneity and consistency and open a space for play and creativity. Taxonomy becomes, in Mark Dion's practice, a basis for a conversation;[31] for Ogunbiyi, it is an opportunity for reframing. As time capsules made possible by techniques of suspended, delayed, or assisted animation, collections are past, present, and future. Their temporalities are layered and open; for Derrida, it is this "movement of the promise of the future no less than of recording the past" that gives the archive "an unknowable weight."[32] This unknowable weight may encompass the history of the concept and the history of a collection, but also its future activations. Ogunbiyi's distinctive annotations of the herbarium sheet capture exquisitely, deliberately, this forward-looking promise of the archive: "You Will . . ."

Notes

1 Nicholas Jardine, James A. Secord, and Emma C. Spary, eds., *Cultures of Natural History* (Cambridge: Cambridge University Press, 1996).

2 On the history of herbaria, see Maura C. Flannery, *In the Herbarium: The Hidden World of Collecting and Preserving Plants* (New Haven: Yale University Press, 2023); Barbara M. Thiers, *Herbarium: The Quest to Preserve and Classify the World's Plants* (Portland, Ore.: Timber Press, 2020); Brian W. Ogilvie, *The Science of Describing: Natural History in Renaissance Europe* (Chicago: University of Chicago Press, 2006); and Chiara Nepi, "Botanical Collecting, Herbaria and the Understanding of Nature," in *Changing Perceptions of Nature*, ed. Ian Convery and Peter Davis (Woodbridge: Boydell Press, 2016), 89–98.

3 Nepi, "Botanical Collecting," 89.

4 Daniel S. Park, Xiao Feng, Shinobu Akiyama, et al., "The Colonial Legacy of Herbaria," *Nature Human Behaviour* 7 (2023): 1059–68, https://doi.org/10.1038/s41562-023-01616-7.

5 Michel Foucault, *The Order of Things: An Archaeology of the Human Sciences* (New York: Vintage Books, 1994), 131.

6 Foucault, *The Order of Things*, 134–35.

7 Foucault, *The Order of Things*, 132.

8 Foucault, *The Order of Things*, 137. On the flattening effect of representation, see also Giovanni Aloi, ed., *Why Look at Plants? The Botanical Emergence in Contemporary Art* (Leiden: Brill, 2019).

9 Foucault, *The Order of Things*, 129.

10 Staffan Müller-Wille and Isabelle Charmantier, "Natural History and Information Overload: The Case of Linnaeus," *Studies in History and Philosophy of Science: Part C, Studies in History and Philosophy of Biological and Biomedical Sciences* 43, no. 1 (2012): 4–15. For a different line of exploration, see Geoffrey C. Bowker and Susan Leigh Star, *Sorting Things Out: Classification and Its Consequences* (Cambridge, Mass.: MIT Press, 1999).

11 Bruno Latour, *Science in Action: How to Follow Scientists and Engineers Through Society* (Cambridge, Mass.: Harvard University Press, 1987).

12 Lorraine Daston, "Type Specimens and Scientific Memory," *Critical Inquiry* 31, no. 1 (2004): 177.

13 For an interdisciplinary exploration of plant extinction through the lens of collections, and with particular attention to the herbarium, see the special issue of *Arnoldia*, "Collections, Extinction, Abundance," ed. Yota Batsaki and Peter Crane (June 2024).

14 Yota Batsaki, "The Apocalyptic Herbarium: Mourning and Transformation in Anselm Kiefer's *Secret of the Ferns*," *Environmental Humanities* 13, no. 2 (2021): 391–413, https://doi.org/10.1215/22011919-9320211.

15 So far, I have been unable to find the origins of these dimensions. They vary on occasion. I have been told that the Vienna herbarium is slightly different, and that in the past the Natural History Museum, London, used 17.5 × 11.25 in sheets; in the nineteenth century, that discrepancy was used as an argument against a plan to unite its herbarium with the one at Kew by moving everything to London (Maura C. Flannery, personal communication).

16 Temitayo Ogunbiyi, personal communication.

17 Londa Schiebinger, "Taxonomy for Human Beings," in *The Gendered Cyborg: A Reader* (London: Routledge, 2000), 11–37.

18 Yota Batsaki, "The Plant-Human Hybrid Is the Cyborg of Our Moment," *Hyperallergic*, September 7, 2022, https://hyperallergic.com/758951/the-plant-human-hybrid-is-the-cyborg-of-our-moment/.

19 Aloi, *Why Look at Plants?*, 5.

20 See "African Foods in Diaspora: Sustenance, Worldmaking, and Neotropical Legacies," by Judith A. Carney and Richard N. Rosomoff (this volume).

21 Allison Caban and Malek Charchour, "Okra: An Enduring Kitchen Staple," https://lab.plant-humanities.org/okra/.

22 For oral traditions about the transportation of rice in the hair of abducted children among Maroon societies in Brazil, see Judith A. Carney, "'With Grains in Her Hair': Rice in Colonial Brazil," *Slavery & Abolition* 25, no. 1 (2004): 1–27, https://doi.org/10.1080/0144039042000220900.

23 Foucault, *The Order of Things*, 130.

24 Foucault, *The Order of Things*, 159.

25 Temitayo Ogunbiyi, personal communication.

26 Jacques Derrida, *Archive Fever: A Freudian Impression*, trans. Eric Prenowitz (Chicago: University of Chicago Press, 1996), 2.

27 Derrida, *Archive Fever*, 94.

28 Derrida, *Archive Fever*, 12.

29 Banu Subramaniam, *Botany of Empire: Plant Worlds and the Scientific Legacies of Colonialism* (Seattle: University of Washington Press, 2024), 50.

30 Charles C. Davis, "The Herbarium of the Future," *Trends in Ecology & Evolution* 38, no. 5 (2023): 412–23. See also J. Mason Heberling and Bonnie L. Isaac, "Herbarium Specimens as Exaptations: New Uses for Old Collections," *American Journal of Botany* 104, no. 7 (2017): 963–65, https://doi.org/10.3732/ajb.1700125.

31 Ruth Erickson et al., *Mark Dion: Misadventures of a 21st-Century Naturalist* (Boston: Institute of Contemporary Art; and New Haven: Yale University Press, 2017).

32 Derrida, *Archive Fever*, 29.

Poetics

10

THE QUIET SCIENCE?

Jorge Luis Borges's Anti-Botany

LUCAS MERTEHIKIAN

Argentine writer Jorge Luis Borges (1899–1986) did not care much for plants. While his interest in animals is well-documented in many of his most salient short stories, poems, and essays, plants did not receive the same kind of attention from Borges or his scholarly readers. Even when occasionally praising the study of the vegetal world, Borges does so in a rather condescending manner. In an entry in his friend Adolfo Bioy Casares's journal, dated July 14, 1967, Borges is recorded discussing potential titles for his friend's forthcoming novel. One of the options Borges fancies (*La piedra y el árbol* [*The Stone and the Tree*]) elicits from him a brief reflection on the study of plants: "Botany is quieter than zoology. In the zoological garden you must pay admission; if they charged admission to the botanical garden, it would be empty."[1] Not surprisingly, Borges never dedicated a book to plants such as his *Manual de zoología fantástica* (*The Book of Imaginary Beings*) (1957), which describes human encounters with animals as a metaphysical riddle.[2]

Contrarily, Borges's description of botany—and by extension, plants—as "quiet" reproduces common assumptions about the study of plants in Western culture that can be traced back to at least Jean-Jacques Rousseau, who, toward the end of his life, embraced botany as a "salutary science" that provided him refuge from the convulsed world of politics.[3] Like Rousseau, professional naturalists also contributed to this long-standing image of botany as a peaceful and solitary retreat. Although botanical exploration was intimately tied with economics, in their travel writings, naturalists fashioned themselves as merely herborizers, "armed with nothing more than a collector's bag, a notebook, and some specimen bottles, desiring nothing more than a few peaceful hours alone with the bugs and flowers."[4] For literary scholar Mary Louis Pratt, it was this bias that allowed Carl Linnaeus (1707–1778) to present his groundbreaking binomial nomenclature, which rearranged Western knowledge of all existing plants according to their sexual system, as an "abstract, unheroic gesture, with very little at stake," while paving the way for global resource extraction.[5]

The idea of plants as quiet objects of study that invite soothing examination is not at odds with their allegedly uncharismatic nature. Indeed, we often take plants for granted. James H. Wandersee and Elisabeth E. Schussler coined the term "plant blindness" to refer to this "inability to see or notice the plants in one's environment," as humans, consciously or not, tend to rank "plants as inferior to animals and thus, as unworthy of consideration."[6] Those affected by this condition see plants as sessile and motionless, nothing more than the greenery background of animal life. Thus, we fail to notice plants in our daily lives, disregard their centrality in key biochemical cycles, ignore their biological features and cultural histories, and lack awareness of the different timescales of plant activity.[7] In other words, we do not see in them the kind of more-than-human mystery that Borges saw in animal life.

Scholars such as Michael Marder and Jeffrey T. Nealon have claimed that this inattention to plants can also be traced in Western philosophy and culture, with plants populating "the margin of the margin, the zone of absolute obscurity undetectable on the radars of our conceptualities."[8] Nealon argues that while plants make up more than 80 percent of the planet's biomass, they have played the "role of the abjected other . . . throughout the biopolitical era."[9] Too strange to us, lacking faces and a centralized nervous system, plants have been relegated to the lowest ranks of being since Aristotle, who considered that their vegetal soul did not partake in key aspects of life, such as locomotion or perception.[10] The reverse side of the stillness of plants is their seemingly infinite appetite for growth. Thus, plants can be uncanny—familiar but also radically different from us; quiet, but capable of expanding indefinitely. The question then, to a thinker like Marder, is: "How is it possible for us to encounter plants? And how can we maintain and nurture, without fetishizing it, their otherness in the course of this encounter?"[11]

This chapter takes Borges as a case study to answer this question and contribute to ongoing scholarly efforts in the plant humanities to undo this persistent inattention to plants. Through a close reading of one of his most famous short stories, "Funes el memorioso" ("Funes, His Memory") (1942),[12] I propose that literature allows us to glimpse into vegetal, nonhuman ways of being and feeling that would otherwise remain unimaginable. In doing so, literature can enter into conversation with Western scientific and cultural understandings of plants.

Here, I follow in the footsteps of critics such as Patrícia Vieira and Claudette Sartiliot to shy away from reading plants as symbols or mere ornaments. Vieira has coined the term "phytographia" to refer to the "porous boundary between artistic portrayals of flora and the imprints left in texts by the plant themselves."[13] If science examines "plant modes of inscription in the biosphere," then literary studies can analyze the "modes in which the vegetal world is embedded in human culture."[14] Similarly, in her discussion of literary representations of flowers in canonical twentieth-century literature, Sartiliot argues that flowers (and plants more generally) "have no *topos*, no clear or real place."[15] They are often associated with beauty and innocence, but they also represent evil and sin, even within the same text. Instead of tracing these shifting metaphors, I will consider the transformative potential of literature in changing our perception of the vegetal world as sessile and quiet.

Although plants feature scarcely in Borges's literature, there is more than one vegetal entry point to it. The most salient is to consider his lyrical descriptions of roses, which add to a longstanding tradition that Borges links to prestigious names such as that of Giambattista Marino and John Milton.[16] From a more biographical standpoint, one could also choose to trace in his texts his declared love for eucalyptus trees.[17] While I expect to investigate these possibilities in the future, I focus in this chapter on reading plants in connection with a key philosophical problem throughout Borges's work—in his own words, how to grapple with the fact that "there is no classification of the universe that is not arbitrary and conjectural."[18] Thus, in the first section, I present an overview of how plants feature (and are blatantly ignored) in some of Borges's canonical texts addressing this question, specifically "Tlön, Uqbar, Orbis Tertius" (1940), "El Aleph" ("The Aleph") (1949), and "El idioma analítico de John Wilkins" ("The Analytical Language of John Wilkins") (1942). Following this vegetal thread, the remaining sections of this chapter offer a close reading of "Funes, His Memory."

The story's plot is simple: after falling off a horse and being left paralyzed, Ireneo Funes, a young man living in a small village in Uruguay, awakens to an overwhelmingly prodigious sense of perception that allows him to see and feel the world in the utmost detail, to perceive all changes around him, no matter how tiny or negligible, and to remember them. Like the texts mentioned before, this story has also been traditionally read as a "fictional mise-en-scène of the enslavement of a discourse by direct experience"[19] and a foremost example of how Borges's "heterogeneous enumerations and overcrowded series" unveil the artificial nature of classificatory principles.[20] More recently, Funes's infinite memory has been compared to that of artificial intelligence, and the short story has been read as both a precursor to and a critique of posthuman ideas, with Funes embodying a kind of cyborg *avant la lettre*.[21]

My reading builds on these interpretations but foregrounds what remains absent in all of them, despite lying right there in plain sight from the very beginning of the story: plants. I highlight the many instances in which the narrator either portrays Funes looking at plants or uses plants to draw analogies that allow the reader to understand how Funes's abnormal memory works. I propose to read it as an experiment in what I call "anti-botany," to think about it in relation to Michel Foucault's description of botany as the key science in shaping the "age-old distinction between the Same and the Other" on which Western thinking is grounded.[22] Funes, I argue, can be understood as the antithesis of Linnaeus, for whom plants should be reduced to ideal types to be studied; as stated in his *Philosophia botanica* (1751), "the foundation of botany is two-fold, *arrangement* and *nomenclature*."[23] After the accident that makes him seem extremely quiet in the eyes of other people, Funes's mind and body are, in fact, perfectly in tune with imperceptible changes in the environment around him. Funes offers the narrator and the readers a window into a vibrant, wild vegetal world, where thinking might be something other than human. Doing so shows us that the field of plant humanities can illuminate scientific understandings of botany that consider plants in all their liveliness and that have often taken a back seat to the grand narrative of Linnaean taxonomy.

This is indeed a timely reading of Borges's work, as plants have increasingly become an object of artistic and humanistic inquiry in recent years. This plant turn within contemporary literature is perhaps best expressed by the critical and commercial success of books such as Robin Kimmerer's *Braiding Sweetgrass* and Richard Powers's *The Overstory*, which intertwine botany and literature (and, in the case of *Braiding Sweetgrass*, Indigenous knowledge) to offer a bewildering image of plant life that focuses on plant behavior and communication, which, as Kimmerer explains, plant science is only now beginning to unpack. One prominent example that she offers is how pecan trees communicate to support each other and coordinate fruiting—a tenet of Potawatomi botanical knowledge—through pheromones carried by the wind and underground networks of fungi that inhabit tree roots.[24]

Ecocritic Rob Nixon has attributed the success of such books to the fact that readers find in them a cooperative template for society that counters neoliberal individualism, often fueled by biological metaphors such as the "selfish gene."[25] Plants then offer an alternative behavior model that is especially appealing in times of ecological and political turmoil. But while Nixon's insightful hypothesis holds remarkable explanatory power, I argue here that in its encounter with the vegetal world, literature can also be a window into forms of plant life and plant thinking that are so radically different that they resist anthropomorphizing. Thus, a short story like "Funes, His Memory" can estrange our perception of both plants and ourselves. Borges's conviction that the "aesthetic act" is "that imminence of a revelation that is yet not produced"[26] acquires an ecological dimension that is particularly relevant now that climate catastrophe is not imminent but has already happened.

Absent Plants

Borges's claim that "reality is not verbal" reads throughout his most famous texts, which critic Beatriz Sarlo deems perfect examples of "philosophical narrative situations"[27] that seek to disrupt the relationship between words and things while underlining that such a relationship is inevitable. Sarlo explains that these are not narratives in which such philosophical ideas are related by the characters or presented to the reader for consideration but rather that "ideas are the very stuff of the plot, and they shape it from inside."[28] Once again, Borges's philosophical narrative situations ask, if reality is not verbal, how can literature represent it? Borges's philosophical narrative situations complicate realist and nominalist binarism and push the boundaries of language to the edge of chaos. As a result, fictional encyclopedias, dictionaries, libraries, classification systems, and even imaginary worlds proliferate in Borges's literature, especially in texts that resort to what Sylvia Molloy has called "heteronomous enumerations"[29] to represent verbally that which is essentially not verbal—reality. In this section, I review how plants feature in three of Borges's most salient philosophical-literary situations. I show how plants are either omitted or only obliquely present in the heteronomous enumerations in "Tlön, Uqbar, Orbis Tertius," "The Analytical Language of John Wilkins," and "The Aleph" to argue that although relegated to the quiet margins of Borges's literature (and sometimes even erased), plants can be a productive entry point into the question of whether literature can bring representational order to an inherently chaotic world.

Much has been written about how "Tlön, Uqbar, Orbis Tertius" imagines an alternative reality that dangerously begins to spill over into our own. The story narrates the unlikely finding of a volume belonging to a mysterious encyclopedia created by a secret society in the nineteenth century. *A First Encyclopedia of Tlön* gives abundant information on this imaginary world, where endless philosophies are created successively because, in Tlön, the narrator explains, metaphysics is a kind of literature. Sarlo calls these theories "*as if* philosophies,"[30] underlining the fictional quality of all reasoning in Tlön. The world is a series of unique acts and, therefore, every act of naming is seen as a form of falsification. This explains why there are no nouns in Tlön's language and why identity is not a guiding principle of Tlön's philosophies.

When the narrator first finds a volume of Tlön's encyclopedia, he briefly summarizes its content. To do so, he crafts a heteronomous enumeration that brings together disjointed elements and anticipates the discordant philosophies of Tlön: "I now held in my hands a vast and systematic fragment of the entire history of an unknown planet, with its architecture and its playing cards, the horror of its mythologies and the murmur of its tongues, its emperors and its seas, its minerals and its birds and its fishes, its algebra and its fire, with its theological and metaphysical controversies."[31] In Tlön, there are not only philosophical controversies and political figures constructs but also organic and inorganic matter. There are birds and fish, and even rocks, seas, and fire. However, there seem to be no plants, no vegetal life whatsoever. Plants remain unseen—or rather, unimagined. It is probably not relevant to interrogate this absence from a biological point of view (how would life be supported in a world without plants?). Still, I believe it is productive to consider it on two different levels. On the one hand, the absence might be telling of the philosophical dismissal of plants that thinkers such as Marder and Nealon have examined. On the other hand, it hints at a way of thinking about plants that makes for common understandings of Western botany since the eighteenth century. If, as historian of science Lorraine Daston writes, "botanists are those who know how to name all the plants,"[32] then it is only natural that there are no botanists in Tlön, where the very act of naming is questionable, and that there are no plants either, since they would only be known through naming and classification. Consider, in this sense, the botanical importance of "type specimens." Daston explains that to undo the chaos of synonymy and scientific amnesia that existed since the nineteenth century, botanists decided to attach a herbarium specimen to each species name: each time a botanist names a new species, a single specimen is stored as a future reference for every other botanist who may want to consult it before naming a new species. Thus, "the singularity of the type specimen safeguards the permanence of the name."[33] Contrarily, Borges writes, "things duplicate themselves in Tlön; they also tend to grow vague or 'sketchy,' and to lose detail when they begin to be forgotten."[34] If there were a science of plants in Tlön, it would certainly not rely on type specimens as defined by Daston.

Plants are equally neglected in "The Analytical Language of John Wilkins," an essay where Borges discusses the possibility (or rather the impossibility) of establishing guiding principles for a universal language. After reviewing Wilkins's failed experiment to create such a system in the seventeenth century, Borges claims to know about a Chinese encyclopedia entitled *Celestial Emporium of Benevolent Knowledge* that attempted something similar: "On those remote pages it is written that animals are divided into (a) those that belong to the Emperor, (b) embalmed ones, (c) those that are trained, (d) suckling pigs, (e) mermaids, (f) fabulous ones, (g) stray dogs, (h) included in the present classification, (i) frenzied, (j) innumerable, (k) drawn with a very fine camel hair brush, (l) etcetera, (m) having just broken the water pitcher, (n) that from a long way off look like flies."[35]

This is the passage that famously moved Foucault to laughter and inspired his book *The Order of Things*, grappling with the fact that "the monstrous quality that runs through Borges's enumeration . . . consists in the fact that the common ground on which such meetings are possible has itself been destroyed."[36] However, it is worth noting that, once again, plants take a back seat to animals, further reinforcing Borges's claim that plants are too quiet to move readers to wonder and laughter and even raising the point that however "arbitrary and conjectural" classification systems might be, they do seem to follow certain patterns in privileging some forms of life over others.

This pattern reappears in a somewhat different guise in "The Aleph," arguably Borges's most famous text. The story narrates the discovery of a prodigy: a point in a staircase, in the basement of a house in Buenos Aires, through which the whole universe can be seen *simultaneously*. The Aleph is less an object than a vantage point that allows the viewer to see everything there is, was, and will be in the entire universe, which makes the Aleph impossible to render in writing. As he sets out to explain what he saw when he stared into it, the narrator is confronted with the representational conundrum that animates Borges's narrative philosophical situations—how to order chaos? To highlight this impossibility, rather than mending it, the narrator offers yet another heteronomous enumeration that runs for many lines, featuring animals ("horses with flowing manes"; "tigers, pistons, bisons"), inorganic elements ("tides"), and insects ("all the ants on the planet"), among many other *disjecta membra* that make up the series.[37] The Aleph challenges the possibility of describing the world as an orderly arrangement of objects and beings. Moreover, ecocritic David Farrier argues that today, the enumeration prompted by the Aleph can be read as an aesthetic response to the Anthropocene, which demands that we derange traditional spatial and temporal scales to grasp the geological nature of the climate crisis.[38]

But how do plants feature in this rowdy vision of the universe? The only explicit mentions of plants seen through the Aleph are not as detailed as those of horses or ants ("a bunch of grapes"; "tobacco"). Still, plants are also mentioned obliquely as reflection ("I saw a sunset in Querétaro that seemed to reflect the color of a rose in Bengal"), as shadows ("I saw the slanting shadows of ferns on a greenhouse floor"), and as absent ("I saw a ring of baked mud in a sidewalk, where before there had been a tree"). I want to argue that these elusive references to plants underline their familiarity and strangeness and pose another problem in literary representation. Can the life of plants be represented, or are they so opaque, so *quiet*, that they can only manifest indirectly? This is, I will argue, the theme of "Funes, His Memory"—how the literary representations of plants might illuminate other-than-human (and potentially richer) forms of life. My claim is that plants are key in the narrative to understanding Funes's exceptional perception and memory and, therefore, to speculating about this question.[39]

The Vegetal World of Funes

"Funes, His Memory" is a straightforward narrative: after a horse accident in Fray Bentos, a provincial town in Uruguay, the young Ireneo

Funes awakens with prodigious perception and memory. The story is told from the point of view of an Argentinean (named Borges) who meets Funes while visiting Uruguay, two years before the accident. Back then, Funes already displayed an abnormal skill—he told the time at any given moment without consulting a clock. The narrator remembers this feat when he returns to Uruguay, only to learn that Funes has fallen off a horse and been left paralyzed. "I was told", the narrator explains, "that Ireneo *never stirred* from his cot, with his eyes fixed on a *fig tree* behind the house or on a spiderweb"[40] The narrator catches a glimpse of Funes in this state and feels sorry for him: "Twice I saw him behind the iron grate that crudely underscored his prisoner-like state: once lying *motionless*, with his eyes closed; the second time *motionless as well*, absorbed in the *contemplation of a fragrant switch of artemisia.*"[41]

The insistence on the fact that Funes does not move is especially relevant because now that Funes is condemned to stillness, whenever he has his eyes open, he looks precisely at plants, whether a fig tree in the back of the house or a sprig of artemisia ("santonica," in the Spanish original; it is a plant used in South American folk medicine as a remedy for roundworms). But what Funes sees is not at all quiet, as we usually tend to think of plants. We will soon learn from the narrator that what Funes sees is full of life and ceaseless movement.

Before leaving town, the narrator pays his friend a visit to collect a Latin dictionary and a volume of Pliny's *Natural History* that he had loaned to him.[42] When they meet, he witnesses a prodigy like the existence of the Aleph. He explains:

> When [Funes] fell, he'd been knocked unconscious; when he came to, the present was so rich, so clear, that it was almost unbearable, as were his most trivial memories . . . He reasoned (*or felt*) that *immobility* was a small price to pay. Now his perception and his memory were perfect.
>
> With one quick look, you and I perceive three wineglasses on a table; Funes perceived every grape that had been pressed into the wine *and all the stalks and tendrils of its grapevine.*"[43]

There is much to unpack in this passage. First, it should be noted that the narrator insists once again on Funes's immobility. But now we learn more about what is really going on in his world. To Funes, every moment is an Aleph, and because memories are just as vivid as the present, the awareness of time that would make humans superior is now deranged. Once again, the narrator's problem is how to put this into words and how to make literature out of it.[44]

But, unlike in the short stories examined above, here plants might be the key to answering this question. The "grapes" that the narrator sees in the Aleph now have expanded into leaves, tendrils, and fruit. In other words, the narrator in "Funes, His Memory" resorts to plants, instead of eschewing them, to explain that which in his narrative is unexplainable. Plants are a kind of exemplary model to understand Funes's prowess. It is assumed that they are so intricate, complex, and ever-changing that to be *fully* attentive to them, one should be like Funes. Confronted with the "most difficult" part of the story he is telling, the narrator resorts to plants once again to try to grasp Funes's unimaginable world.

The parentheses also convey important information—Funes did not reason that immobility was a low price to pay but rather *felt* it. Although the narrative tends to stress Funes's sense of sight (he stares at the fig, he sees the grapes), his perception is not limited to just one sense. Later, the narrator explains that Funes's memories (which is to say, his present) are not simply images or even moving images. They are linked to "muscular sensations, thermal sensations, etcetera." Once more, references to plants are key in this sense, as we can now reread the passage in which Funes is described "contemplating" a *fragrant* spring of artemisia. Contemplation should not be understood here as a synonym for sight, and that is why, quoting the narrator, Funes *perceives* and does not just "see" or "look" at things. Contemplation for Funes is a synesthetic term that conveys all senses without privileging any and reminds us that every time Funes perceives something, he does not only look at it but bodily experiences it in a way that is so vast that all words become insufficient—therefore, the narrator's "etcetera," which we cannot even fathom what it contains in Funes's world.

Funes is certainly not "plant blind," and neither is the narrator, despite Borges's disregard for botanical gardens. We know that Funes spent his days "with his eyes fixed on a fig tree," which most people interpreted as a sign of his physical decay. But now we also know that Funes must have been contemplating an exhilarating spectacle of vegetal life, with its movement, growth, retraction, and even death, each instant filled with its distinctive smell, heat, "etcetera," all equally unforgettable. It is clear by now that the mention of plants is almost tantamount to the mention of Funes's extraordinary abilities. In fact, plants are already present in the short story's opening sentence, in which the narrator first introduces (without naming) Ireneo Funes: "I recall him . . . holding a dark *passionflower* in his hand, seeing as it had never been seen, even had it been stared at from the first light of dawn till the last light of evening for an entire lifetime."[45]

Surprisingly, although being in such a prominent position in the story, the passionflower (*Passiflora*) has gone unnoticed. The choice is interesting because the passionflower has a long-standing cultural history in the West as a symbol of martyrdom, as it was largely used in the Americas during colonial times to preach Christianity. In these teachings, the ten petals represent the ten faithful apostles; the top three stigmas stand for the three nails on the cross; the lower five anthers are Christ's five wounds; and the flower's fragrance is reminiscent of the spices prepared by the Holy Women.[46] Furthermore, in Argentina, the *pasionaria* was the first flower to be nominated by a committee of botanists as the country's national flower in 1914, a rank that would be later officially awarded to the "ceibo" (*Erythrina crista-galli*) by a 1940 executive order.[47]

Sartiliot argues that as metaphors, flowers "have no *topos*, no clear or real place," meaning that their metaphorical reference can change from life and innocence to evil and sin for no apparent reason.[48] Similarly, Jacques Derrida sees flowers as "the zero degree in the chain of signification."[49] In literature, Sartiliot concludes, "flowers appear only to disappear, are present only as they metamorphose themselves endlessly into other things."[50] For Funes, this endless transformation is a *literal* rather than metaphorical process: he can see and feel the passionflower he holds in his hand changing dramatically from one moment to another, in a way that the narrator cannot put into words. Differences are too notorious for Funes to extrapolate from the passionflower any kind of symbolism, to see it as a symbol of martyrdom or a botanical representation of nationhood—and this is precisely why I find it extremely interesting that Funes is holding a flower with such a loaded cultural history. In other words, it is not that Funes's passionflower represents "the zero degree in the chain of signification," but that it challenges any possibility of signification at all. Paraphrasing Gertrude Stein, a passionflower is a passionflower is a passionflower. Funes's extraordinary perception makes the very naming of the flower so unstable that language becomes as chimeric as in Borges's fictional Chinese encyclopedia of animals. If no two things are equal to each other, nothing can be named. It would be more accurate, perhaps, to say that, to Funes, a passionflower is not a passionflower and cannot be a passionflower. There might not be, in fact, such a thing as a passionflower.

Funes as the Anti-Linnaeus

Daston has described botany as the science of naming and classification. If that is the case, then Funes could be described as an anti-botanist of sorts. Since Linnaeus's invention of the binomial nomenclature in the eighteenth century, Daston writes, "botanists are those who know how to name all the plants."[51] Naming, as a form of stabilizing the study of plants, would be a way to remedy the absurdity of Borges's fictional Chinese encyclopedia, to arrange the chaos of the Aleph, and to constrain Funes's expanding perception. According to Daston and Peter Galison, Linnaeus's naming enterprise relied on "the ability to synthesize impressions"[52] and to extract "the typical" from the particularities of nature. This is what Daston and Galison call the scientific model of "truth-to-nature." Truth-to-nature demanded from botanists a way of looking at, depicting, and classifying plants that was "aggressively selective"[53] in telling apart that which is necessary to name a species from that which is not.

This epistemological principle is shown in Linnaeus's definition of how representations of plants should be constructed for the stabilization and study of the vegetal world. A botanical illustration, he explains in *Philosophia botanica*, "represents the plant as if it were in a mirror," although but a mirror filtering out all physical features except "Number, Shape, Position, and Proportion."[54] Consider an example from Linnaeus's *Hortus cliffortianus* (1737), showing types of leaves for classification purposes. Here, plants certainly appear to be "quiet" (Figure 10.1).

This kind of simplified perception for classification purposes is completely alien to Funes's world. Indeed, how could Funes reduce his overwhelmingly rich experience of plant life to a set of schematic drawings made to reduce all differences to just the *important* ones? "The naturalist," Foucault says in *The Order of Things*, is "concerned with the structure of the visible world and its denomination according to its characters. Not with Life."[55] It is difficult not to consider "the naturalist" (i.e., Linnaeus) as Funes's antagonist. The narrator gives a brief but powerful account of how Funes's impossible natural history system would work: "In the seventeenth century, Locke postulated (and condemned) an impossible language in which each *individual thing—every stone, every bird, every branch*—would have its own name; Funes once contemplated a similar language but discarded the idea as too general to him, too ambiguous. The truth was, *Funes remembered not only every leaf of every tree in every patch of forest, but every time he had perceived or imagined that leaf.*"[56]

Unlike in the heterogeneous enumerations that we analyzed above, elements such as "stones and birds" take a back seat to plants in this passage, and the narrator resorts again to vegetal life to convey Funes's extraordinary sense of perception and memory. If looked at fully (like only Funes can), not two leaves are alike; not even one leaf is identical to itself. To Funes, the growth of plants, which to most of us is invisible and oftentimes sudden, is just a continuous progression in which every instant remains sharply distinctive, so much so that it would be impossible to extract from them the kind of features that, to Linnaeus, were essential to classify the natural world. Whereas for the narrator, once and again, plants are the utmost example of instability and change, a kind of paradoxical *type* that shows that there are no types. Furthermore, a classification system that accounts for Funes's world would not only include that which is perceived and memorized but also that which is imagined.

Linnaean taxonomy demanded that botanists master the art of forgetting, which is precisely what Funes cannot do. They had to forget all the variable aspects of plants that could not be rendered in schematic descriptions, drawings, and lifeless herbarium specimens. In *Philosophia botanica*, Linnaeus specifically claims that "the specific name ought to be derived from parts of the plant *that do not vary*."[57] He lists those features that are not essential to plant description and, therefore, should be disregarded when naming and classifying new species, including color ("*Colour* within the same species is remarkably sportive, and so is of no value in definitions"[58]); scent ("*Scent* never clearly distinguishes a species . . . Scents do not allow for fixed boundaries and cannot be defined"[59]), taste ("*Taste* is often variable, according to the person doing the testing"[60]), and potency ("*Potency* and *use* provide definitions that are worthless to a botanist"[61]). Funes's relation with plants is defined by the narrator using adjectives that engage a sensorial experience in sharp opposition to these Linnaean precepts. Funes holds a "dark passionflower" and contemplates a "fragrant swatch of artemisia," and each of his memories is connected to bodily sensations that exceed sight, the privileged sense of natural history.[62] Unlike Linnaeus, Funes cannot but indulge in sensorial experiences without fixed boundaries. Thus, the narrator suspects his ability to think. "To think," he writes, "is to forget differences, to generalize, to abstract. In the teeming world of Funes, there was nothing but particulars—and they were virtually immediate particulars."[63] But what if Funes's perception points instead to other ways of thinking about the plant world? What vegetal narratives might his inability to classify illuminate, and how do they intersect with contemporary representations of plant life in literature and culture?

FIGURE 10.1
Folia simplicial, from Carl Linnaeus, *Hortus cliffortianus* (Amsterdam, 1737). Photograph courtesy of Dumbarton Oaks, Washington, D.C.

Conclusion

Although narratives of botany as a taxonomic science have been emphasized in historical accounts of its emergence and consolidation since the early modern period,[64] other scientific traditions ran parallel to it, highlighting instead a plant world of action and movement more akin to Funes's unbound sense of perception than to his alleged lack of thought. In this section, I want to argue that a close reading of "Funes, His Memory" can force us to redirect our attention toward these stories, which are being reactivated today through both cultural and scientific debates.

In opposing taxonomic and behavioral approaches to plant life, I am aware that this is only an apparent contradiction—Linnaeus's contributions to ecology exceed this narrow view of botany as a taxonomic science, as I will mention later, and plant taxonomies can be the very basis for further environmental knowledge and action. My interest here lies in the ways in which Linnaean botany depended on taxonomic distinctions and how these have come to be understood by historians and cultural critics such as Daston, Foucault, and Pratt (to name but a few) as the tenets of a "quiet" science. In his *Essay on the Geography of Plants* (1805), Alexander von Humboldt seems to trace two trends in botany in the early nineteenth century that might shed light on the distinction that I try to make here: "Botanists usually direct their research towards objects that encompass only a very small part of their science. They are concerned almost exclusively with the discovery of new species of plants, the study of their external structure, their distinguishing characteristics, and the analogies that group them together into classes and families. The knowledge of the forms which make up organized beings is no doubt the principal basis for descriptive natural history... [But] it is no less important to understand the geography of plants... This science, as vast as its object, paints with a broad brush the immense space occupied by plants."[65]

Such interest in plant communities and their associated behavior predates Humboldt, of course. In fact, a decade before Linnaeus published his *Species plantarum* (1753), where he applied the binomial nomenclature consistently for the first time, the French naturalist Georges-Louis Leclerc de Buffon wrote a *Mémoire sur la culture des forêts* that, according to American botanist Frederic Clements, contains "the first clear record of the succession of forest dominants, and of the effect of light and shelter on the process."[66] Buffon noted that the shrubs and bushes that could resist wind and frost became shelters for young trees, paving the way for them and allowing them to grow, eventually dominating the landscape and suppressing plant life around them.[67] As Frank N. Egerton explains, Linnaeus himself advanced the study of plant succession "from practically bare rock with lichens through a series of plant stages culminating in a forest followed by death and decay of trees and the use of decaying wood by fungi and beetles to create a revolving cycle of nutrients."[68] It is interesting to note how this essay (translated from Swedish into Latin as *Stationes plantarum* in 1754) foregrounding plant life rather than plant naming has not received the same scholarly attention from historians of science.

Rosetta S. Elkin has noted that botanical traditions that consider plant life from microcosmic activity to their relationship with the environment have been overshadowed by taxonomy defining plants as stable artifacts represented by idealized type specimens.[69] We find reactions against such trends already in nineteenth-century sources, as mentioned above. In the prologue to his *Essay on the Geography of Plants*, Humboldt privileged the study of plant association due to their physiognomy, whereas "descriptive botany classifies plants according to the resemblance of their smallest but most essential parts, those relating to fructification."[70] Moreover, he disregarded the utility of botanical collections in this line of research, writing: "An artist of distinction would find it worthwhile to study, not in greenhouses or in botany books but in nature itself, the physiognomy of the plant groups that I have enumerated."[71] To Humboldt, plant life had to be observed without mediation, that is, in its ever-changing state that Funes cannot help perceiving.

The study of plant life through movement rather than fixed categories is probably best exemplified by Charles Darwin's studies. Although not nearly as well-known as *On the Origin of Species*,

Darwin published six books on plant behavior, including treatises on orchids and climbing plants. His passion for climbing plants and their behavior, in fact, was prompted by his own inability to move—much like Funes. During one of his periods of sickness, and after having read a paper by American botanist Asa Gray, Darwin became interested in closely observing the development of the tendrils of *Echinocystis*. He noted in a letter to Gray that the plant's leading shoots would rotate incessantly, bringing "the tendrils into contact with any body within a circle of a foot or 20 inches in diameter."[72] Eventually, Darwin's observations would lead to *On the Movements and Habits of Climbing Plants*, where he challenged the common view that "plants are distinguished from animals by not having the power of movement. It should rather be said that plants acquire and display this power only when it is of some advantage to them."[73] In challenging this assumption, Darwin questions common characterizations of plants (and botany) as quiet, from which Borges certainly drew.

A close reading of "Funes, His Memory" that pays attention to the subtle but meaningful ways in which the narrator represents Funes's relationship with the vegetal world can certainly help illuminate these overlooked narratives of plant life and movement. It can also prompt a rather speculative question that looks even deeper into plant behavior—what if we considered Funes's extraordinary sense of perception as more akin to that of plants than humans? Could this be the basis for an exploration of plant thought beyond language? We saw in the introduction that over the past few years, works like Robin Kimmerer's *Braiding Sweetgrass*, Richard Powers's *The Overstory*, Peter Wohlleben's *The Hidden Life of Trees*,[74] and Paco Calvo and Natalie Lawrence's *Planta Sapiens*,[75] to name a few, have gained scholarly and popular attention. While there might be many reasons for this, these books emphasize an understanding of vegetal life that is not at all static and motionless, regardless of appearances, but rather "teeming," as the plant world represented by Buffon's and Linnaeus's studies of ecological succession in the forest, Humboldt's appreciation of plant geography, and Darwin's observations of climbing plants.

Furthering this line of inquiry about plant behavior, the question of whether plants should be deemed intelligent or not remains a controversial one in the emerging field of plant neurobiology. There is, however, a growing scientific and nonscientific interest in redefining what we understand by "plant intelligence." In a 2003 review of recent discoveries on the subject, biologist Anthony Trewavas claimed that "the use of the term 'vegetable' to describe unthinking or brain-dead human beings perhaps indicates the general attitude [toward plant intelligence],"[76] but he also pointed to a consensus that, like animal behavior, "plant development is plastic too and is not irreversible."[77] Regardless of the outcome of this controversy, this growing interest in the question of plant behavior and intelligence reflects a way of thinking about the vegetal world in which, I would argue, "Funes, His Memory" takes part (and even anticipates).

Plant biologists such as Stefano Mancuso have argued that aspects of vegetal intelligence such as learning, memory, and plasticity are beginning to be uncovered, revealing plant life as anything but quiet, let alone still. Rather, it seems to operate on radically different timescales than our own—too fast or too slow. Thus, plants are motionless only to the human eye. As an example of this frenetic but imperceptible plant activity, biologist Monica Gagliano cites the case of shrubs that can catapult their pollen grains at "lightning speeds of about 0.5 meters per second," with an acceleration rate of "about eight hundred times that experienced by astronauts during takeoff" to propagate.[78] Taking the direct observations of someone like Darwin to an extreme, this might be the kind of vertiginous life that Funes sees (and, more importantly, feels) when contemplating the leaves in the wood, the fig tree in his backyard, a sprig of artemisia, the climbing tendrils of a grapevine, and the passionflower in his hand.

Funes's botany is not one focused on classification, but a vitalist, interconnected one that follows in the footsteps of more obscured trends in the study of plant life. Just like it happens with the vegetal world around him, Funes's own motionless, still life is only apparent. His mind and body ("thermal sensations, muscles, etcetera") are, in fact, in perfect tune with his environment. A kind of plant-human hybrid, Funes experiences the world in such a way that every instant is both fleeting and fixed in his bodily memory. Plant-thinking, according to Marder, starts with the explosion of identity. This might mean the end of one kind of thinking (the one that begets both successful and absurd classifications of the world). But Funes's world seems to be the most vibrant of them all, his life a kind of sensorial adventure that the human mind can also glimpse at. To offer a window into radically different forms of life through plants might be the shared task of botany and literature.

Notes

1 "La botánica es más tranquila que la zoología. En el jardín zoológico hay que pagar la entrada; si cobraran la entrada en el botánico, quedaría vacío." Adolfo Bioy Casares, *Borges*, ed. Daniel Martino (Buenos Aires: Destino, 2006), 1218 (translation by the author). Unless otherwise noted, translations into English are from published English versions of Borges's works. I include the Spanish version for further reference, and occasionally to direct the reader's attention to interpretations that might be lost in translation. In this passage, Borges refers to the zoological and botanical gardens in Buenos Aires, Argentina, founded in 1888 and 1898, respectively. The zoological garden was closed to the public in 2016, while the botanical garden remains open and free.

2 "A small child is taken to the zoo for the first time," Borges writes in the preface to the book, where "he sees for the first time the bewildering variety of the animal kingdom, and this spectacle, which might alarm or frighten him, he enjoys . . . How can we explain this everyday and yet mysterious event?" Jorge Luis Borges and Margarita Guerrero, *The Book of Imaginary Beings* (New York: Avon, 1970), 13.

3 Alexandra Cook, *Jean-Jacques Rousseau and Botany: The Salutary Science* (Oxford: Voltaire Foundation, 2012).

4 Mary Louis Pratt, *Imperial Eyes: Travel Writing and Transculturation* (New York: Routledge, 1993), 27.

5 Pratt, *Imperial Eyes*, 33.

6 James H. Wandersee and Elisabeth E. Schussler, "Preventing Plant Blindness," *The American Biology Teacher* 61, no. 2 (February 1999): 84, https://doi.org/10.2307/4450624. The term has been criticized for its ableist undertones. Kathryn M. Parsley has suggested instead the term "plant awareness disparity" (PAD). Kathryn M. Parsley, "Plant Awareness Disparity: A Case for Renaming Plant Blindness," *Plants, People, Planet* 2, no. 6 (October 2020): 598–601, https://doi.org/10.1002/ppp3.10153.

7 Wandersee and Schussler, "Preventing Plant Blindness," 84–85.

8 Michael Marder, *Plant-Thinking: A Philosophy of Vegetal Life* (New York: Columbia University Press, 2013), 2.

9 Jeffrey T. Nealon, *Plant Theory: Biopower and Vegetable Life* (Stanford: Stanford University Press, 2014), 11.

10 Marder, *Plant-Thinking*, 22.

11 Marder, *Plant-Thinking*, 3.

12 Sometimes also translated as "Funes, the Memorious." Here, I quote from Jorge Luis Borges, *Collected Fictions*, trans. Andrew Hurley (London: Penguin Books, 1999).

13 Patrícia Vieira, "*Phytographia*: Literature as Plant Writing," in *The Language of Plants: Science, Philosophy, Literature*, ed. Monica Gagliano, John C. Ryan, and Patrícia Vieira (Minneapolis: University of Minnesota Press, 2017), 218.

14 Vieira, "*Phytographia*," 218.

15 Claudette Sartiliot, *Herbarium Verbarium: The Discourse of Flowers* (Lincoln: University of Nebraska Press, 1993), 17.

16 See Borges's short prose "La rosa amarilla" ("The Yellow Rose"), included in *El hacedor* (*Dreamtigers*) (1960) and his poem "Una rosa y Milton" ("A Rose and Milton"), included in *El otro* (*The Other*) (1964). He also published a poetry collection named after one of the poems, *La rosa profunda* (*The Deep Rose*) (1975). Although I intend to explore these literary representations of roses in the future, here I am more interested in a less symbolic literary approach to plants, which, I argue, can be the foundation of a new reading of "Funes, His Memory."

17 In the preface of his book of poems, *Adrogué* (Adrogué: Ediciones Adrogué, 1977), Borges claims that to him, the smell of eucalyptus will always remind him of his childhood summer house in Adrogué, a suburban town in Buenos Aires (7). Eucalyptus trees are also present in two of his most well-known short stories, "La muerte y la brújula" ("The Death and the Compass") and "El sur" ("The South"), which also riff on autobiographical themes.

18 "No hay clasificación del universo que no sea arbitraria y conjetural." Jorge Luis Borges, *Other Inquisitions, 1937–1952* (Austin: University of Texas Press, 1964), 104; and Jorge Luis Borges, *Obras completas* (Buenos Aires: Editorial Sudamericana, 2014), 2:91.

19 Beatriz Sarlo, *Jorge Luis Borges: A Writer on the Edge*, trans. John King (London: Verso, 1993), 30.

20 Sylvia Molloy, *Signs of Borges*, trans. Oscar Montero (Durham: Duke University Press, 1994), 112.

21 Ivan Callus and Stefan Herbrechter, "Introduction: Did Someone Say 'Cy-Borges'?," in *Cy-Borges: Memories of the Posthuman in the Work of Jorge Luis Borges*, ed. Ivan Callus and Stefan Herbrechter (Plainsboro Township: Associated University Presses, 2009), 22–23.

22 Michel Foucault, *The Order of Things: An Archaeology of the Human Sciences* (New York: Vintage Books, 1994), xv.

23 Carl Linnaeus, *Philosophia botanica*, trans. Stephen Freer (New York: Oxford University Press, 2003), 111 (emphasis in the original).

24 Robin Kimmerer, *Braiding Sweetgrass: Indigenous Wisdom, Scientific Knowledge, and the Teachings of Plants* (Minneapolis: Milkweed Editions, 2013), 20; and Richard Powers, *The Overstory: A Novel* (New York: W. W. Norton, 2018). *The Overstory* includes a fictional character based on biologist Suzanne Simard, who proposed the notion of the "wood wide web" to explain how organisms share resources through mycorrhizal networks. Scientists have recently claimed that there has been a positive bias toward this theory that is disconnected from evidence. Justine Karst, Melanie D. Jones, and Jason D. Hoeksema, "Positive Citation Bias and Overinterpreted Results Lead to Misinformation on Common Mycorrhizal Networks in Forests," *Nature Ecology and Evolution* 7 (2023): 501–11, https://doi.org/10.1038/s41559-023-01986-1.

25 Rob Nixon, "The Less Selfish Gene: Forest Altruism, Neoliberalism, and the Tree of Life," *Environmental Humanities* 13, no. 2 (2021): 348–71, https://doi.org/10.1215/22011919-9320189.

26 Borges, *Other Inquisitions*, 120.

27 Sarlo, *Jorge Luis Borges*, 30.

28 Sarlo, *Jorge Luis Borges*, 30.

29 Molloy, *Signs of Borges*, 112.

30 Sarlo, *Jorge Luis Borges*, 65.

31 "Ahora tenía en las manos un vasto fragmento metódico de la historia total de un planeta desconocido, con sus arquitecturas y sus barajas, con el pavor de sus mitologías y el rumor de sus lenguas, con sus emperadores y sus mares, con sus minerales y sus pájaros y sus peces, con su álgebra y su fuego, con su controversia teológica y metafísica." Borges, *Collected Fictions*, 71–72; and Borges, *Obras completas*, 1: 726.

32 Lorraine Daston, "Type Specimens and Scientific Memory," *Critical Inquiry* 31, no. 1 (Autumn 2004): 165, https://doi.org/10.1086/427306.

33 Daston, "Type Specimens and Scientific Memory," 162.

34 "Las cosas se duplican en Tlön; propenden asimismo a borrarse y a perder los detalles cuando los olvida la gente." Borges, *Collected Fictions*, 78; and Borges, *Obras completas*, 1:732.

35 "En sus remotas páginas está escrito que los animales se dividen en (a) pertenecientes al Emperador, (b) embalsamados, (c) amaestrados, (d) lechones, (e) sirenas, (f) fabulosos, (g) perros sueltos, (h) incluidos en esta clasificación, (i) que se agitan como locos, (j) innumerables, (k) dibujados con un pincel finísimo de pelo de camello, (l) etcétera, (m) que acaban de romper el jarrón, (n) que de lejos parecen moscas." Borges, *Other Inquisitions*, 103; and Borges, *Obras completas*, 2:91.

36 Foucault, *The Order of Things*, xvi.

37 Molloy, *Signs of Borges*, 165. The entire enumeration in the Spanish original reads, "Vi el populoso mar, vi el alba y la tarde, vi las muchedumbres de América, vi una plateada telaraña en el centro de una negra pirámide, vi un laberinto roto (era Londres), vi interminables ojos inmediatos escrutándose en mí como en un espejo, vi todos los espejos del planeta y ninguno me reflejó, vi en un traspatio de la calle Soler las mismas baldosas que hace treinta años vi en el zaguán de una casa en Fray Bentos, vi *racimos*, nieve, *tabaco*, vetas de metal, vapor de agua, vi convexos desiertos ecuatoriales y cada uno de sus granos de arena, vi en Inverness a una mujer que no olvidaré, vi la violenta cabellera, el altivo cuerpo, vi un cáncer en el pecho, *vi un círculo de tierra seca en una vereda, donde antes hubo un árbol*, vi una quinta de Adrogué, un ejemplar de la primera versión inglesa de Plinio, la de Philemon Holland, vi a un tiempo cada letra de cada página (de chico, yo solía maravillarme de que las letras de un volumen cerrado no se mezclaran y perdieran en el decurso de la noche), vi la noche y el día contemporáneo, vi un poniente en Querétaro que parecía *reflejar el color de una rosa en Bengala*, vi mi dormitorio sin nadie, vi en un gabinete de Alkmaar un globo terráqueo entre dos espejos que lo multiplican sin fin, vi caballos de crin arremolinada, en una playa del Mar Caspio en el alba, vi la delicada osatura de una mano, vi a los sobrevivientes de una batalla, enviando tarjetas postales, vi en un escaparate de Mirzapur una baraja española, vi las sombras oblicuas de unos helechos en el suelo de un invernáculo, vi tigres, émbolos, bisontes, marejadas y ejércitos, vi todas las hormigas que hay en la tierra, vi un astrolabio persa, vi en un cajón del escritorio (y la letra me hizo temblar) cartas obscenas, increíbles, precisas, que Beatriz había dirigido a Carlos Argentino, vi un adorado monumento en la Chacarita, vi la reliquia atroz de lo que deliciosamente había sido Beatriz Viterbo, vi la circulación de mi oscura sangre, vi el engranaje del amor y la modificación de la muerte, vi el Aleph, desde todos los puntos, vi en el Aleph la tierra, y en la tierra otra vez el Aleph y en el Aleph la tierra, vi mi cara y mis vísceras, vi tu cara, y sentí vértigo y lloré, porque mis ojos habían visto ese objeto secreto y conjetural, cuyo nombre usurpan los hombres, pero que ningún hombre ha mirado: el inconcebible universe." Borges, *Obras completas*, 1:967. There are at least three elements that link this enumeration to "Funes, His Memory": the reference to Pliny, to Fray Bentos (the Uruguayan town where the plot is set), and the reference to "rácimos" (bunches), which, as we will see, reappears in "Funes, His Memory" to link his prodigious sense of perception to the vegetal world.

38 David Farrier, *Anthropocene Poetics: Deep Time, Sacrifice Zones, and Extinction* (Minneapolis: University of Minnesota Press, 2019).

39 Botanical references also appear in the lesser-known essay "Penúltima versión de la realidad" (1930), never translated into English, where Borges dissects Polish philosopher Alfred Korzybski's *Manhood of Humanity*. There, Borges pushes back against Korzybski's classification of the world into three different dimensions, with humans reigning over plants and animals, and suggests that humans ruling over them is only one way of perceiving and understanding reality.

40 Borges, *Collected Fictions*, 132 (emphasis by the author). "Me dijeron que no se movía del catre, puestos los ojos en la higuera del fondo o en una telaraña." Borges, *Collected Fictions*, 132 (emphasis by the author); and Borges, *Obras completas*, 1:782.

41 "Dos veces lo vi atrás de la reja, que burdamente recalcaba su condición de eterno prisionero: una, inmóvil, con los ojos cerrados; otra, inmóvil también, absorto en la contemplación de un oloroso gajo de santonina." Borges, *Collected Fictions*, 133 (emphasis by the author); and Borges, *Obras completas*, 1:782.

42 The books dedicated to botany in Pliny's *Natural History* are Books XII to XVIII, so most likely not included in the volume that the narrator loans to Funes. There have been different opinions on whether Pliny's contributions to ancient botany are significant and worthy of study or rather derivative from Theophrastus and unoriginal.

43 "Al caer, perdió el conocimiento; cuando lo recobró, el presente era casi intolerable de tan rico y tan nítido, y también las memorias más antiguas y más triviales . . . Razonó (sintió) que la inmovilidad era un precio mínimo. Ahora su memoria y su percepción eran infalibles. Nosotros, de un vistazo, percibimos tres copas en una mesa; Funes, todos vástagos y racimos y frutos que comprende una parra." Borges, *Collected Fictions*, 135 (emphasis by the author); and Borges, *Obras completas*, 1:784. In the Spanish version, there is a wordplay in the first sentence that the English translation renders only partially. Funes "perdió el conocimiento," literally, "he lost knowledge." This is important, as we will later see, because the narrator considers Funes's

new perception and memory somewhere in between feeling and reason—on the *orilla* between the two, one might say.

44 As he is about to retell his encounter with Funes, the narrator uses almost the exact same words that the narrator in "The Aleph" would use more than a decade later. We read in the Spanish original, "Arribo, ahora, al punto más difícil de mi relato"; in "The Aleph," the narrator goes, "Arribo, ahora, al ineffable centro de mi relato." Borges, *Obras completas*, 1:784, 929.

45 "Lo recuerdo . . . con una oscura pasionaria en la mano, viéndola como nadie la ha visto, aunque la mirara desde el crepúsculo del día hasta el de la noche, toda una vida entera." Borges, *Collected Fictions*, 131 (emphasis by the author); and Borges, *Obras completas*, 1: 781.

46 Oitaven 82.

47 There were many initiatives in the first half of the twentieth century to choose a national flower in Argentina. The first recommendation was issued in 1914 by a committee formed by botanists Ángel Gallardo, Eduardo L. Holmberg, Miguel Lillo, Juan A. Domínguez, and Cristóbal M. Hicken. The committee had been appointed in 1910, during the celebration of the one-hundredth anniversary of Argentina's independence. Later, in 1928, a magazine sponsored by a right-wing nationalist group (*Liga Patriótica*) surveyed its readers, and the magnolia came in first. However, the magazine dismissed the readers' choice, as the magnolia is not native to Argentina, unlike the passionflower (Otaiven 82–83). This context, I believe, can add to a further reading of native and non-native plants in Borges's literature, including his love for eucalyptus, another non-native species, imported in the mid-nineteenth century to Argentina (and currently deemed invasive).

48 Sartiliot, *Herbarium Verbarium*, 17.

49 Sartiliot, *Herbarium Verbarium*, 17.

50 Sartiliot, *Herbarium Verbarium*, 17.

51 Daston, "Type Specimens and Scientific Memory," 165.

52 Lorraine Daston and Peter Galison, *Objectivity* (New York: Zone Books, 2007), 58.

53 Daston and Galison, *Objectivity*, 59.

54 Linnaeus, *Philosophia botanica*, 168.

55 Foucault, *The Order of Things*, 161.

56 "Locke, en el siglo XVII, postuló (y reprobó) un idioma imposible en el que cada cosa individual, cada piedra, cada pájaro y cada rama tuviera un nombre propio; Funes proyectó alguna vez un idioma análogo, pero lo desechó por parecerle demasiado general, demasiado ambiguo. En efecto, Funes no sólo recordaba cada hoja de cada árbol de cada monte, sino cada una de las veces que la había percibido o imaginado." Borges, *Collected Fictions*, 136 (emphasis by the author); and Borges, *Obras completas*, 1:785–86.

57 Linnaeus, *Philosophia botanica*, 221 (emphasis in the original).

58 Linnaeus, *Philosophia botanica*, 229.

59 Linnaeus, *Philosophia botanica*, 231.

60 Linnaeus, *Philosophia botanica*, 231.

61 Linnaeus, *Philosophia botanica*, 232.

62 In this sense, Brian W. Ogilvie explains how herbarium specimens as a key research tool of natural history are devoid of many of the living plant's features that define them, including taste, color, and smell. Brian W. Ogilvie, *The Science of Describing: Natural History in Renaissance Europe* (Chicago: University of Chicago Press, 2006), 169.

63 "[Funes] no era muy capaz de pensar. Pensar es olvidar diferencias, es generalizar, abstraer. En el abarrotado mundo de Funes no había sino detalles, casi inmediatos." Borges, *Collected Fictions*, 137; and Borges, *Obras completas*, 1:786. I have slightly edited Hurley's translation, which reads: "To think is to ignore (or forget) differences . . ." Moreover, the Spanish word "abarrotado" can also be translated as "crowded" and has the word "barrote" in its root, meaning a bar to hold something (or someone) tight. Funes is also described as motionless and behind bars in the beginning of the short story. Elsewhere, the narrator describes his world as "vertiginoso." Borges, *Obras completas*, 1:786.

64 Foucault, *The Order of Things*; and Ogilvie, *The Science of Describing*.

65 Alexander von Humboldt and Aimé Bonpland, *Essay on the Geography of Plants*, ed. Stephen T. Jackson, trans. Sylvie Romanowsky (Chicago: University of Chicago Press, 2009), 64.

66 Frederic Clements, quoted in Frank N. Egerton, "History of Ecological Sciences, Part 54: Succession, Community, and Continuum," *The Bulletin of the Ecological Society of America* 96 (2015): 426–74, https://doi.org/10.1890/0012-9623-96.3.426.

67 Egerton, "History of Ecological Sciences."

68 Egerton, "History of Ecological Sciences."

69 Rosetta S. Elkin, *Plant Life: The Entangled Politics of Afforestation* (Minneapolis: University of Minnesota Press, 2022), 30–34.

70 Humboldt and Bonpland, *Essay on the Geography of Plants*, 74.

71 Humboldt and Bonpland, *Essay on the Geography of Plants*, 74.

72 Charles Darwin to Asa Gray, June 26, 1863.

73 Charles Darwin, *On the Movements and Habits of Climbing Plants* (London: Journal of the Linnaean Society, 1865), 117–18.

74 Peter Wohlleben, *The Hidden Life of Trees: How They Feel, What They Communicate*, trans. Jane Billinghurst (London: William Collins, 2017).

75 Paco Calvo and Natalie Lawrence, *Planta Sapiens* (New York: W. W. Norton, 2023).

76 Anthony Trewavas, "Aspects of Plant Intelligence," *Annals of Botany* 92, no. 1 (July 2003): 1, https://doi.org/10.1093/aob/mcg101.

77 Trewavas, "Aspects of Plant Intelligence," 2.

78 Monica Gagliano, John C. Ryan, and Patrícia Vieira, introduction to *The Green Thread: Dialogues with the Vegetal World* (Lanham, Md.: Lexington Books, 2015), xi, citing Joan Edwards, Dwight Whitaker, Sarah Klionsky, and Marta J. Laskowski, "A Record-Breaking Pollen Catapult," *Nature* 435, no. 164 (2005), https://doi.org/10.1038/435164a.

11

"PHOOLER BHASHA"

Rabindranath Tagore's Flower Linguistics

SUMANA ROY

> There are professors of sciences bordering on the mystical, who declare that they can discriminate the character and disposition of anyone, simply by a careful observation of his handwriting. As to the authenticity of such claims, skepticism is permissible: but there is no doubt that one's handwriting may be modified profoundly by conditions, physical and mental. . . . Such, then, is the history that may be unfolded to the critical eye by the lines and curves of a human autograph. Under a placid exterior, there is also a hidden history in the life of the plant. Storm and sunshine, warmth of summer and frost of winter, drought and rain, all these and many more come and go about the plant. What coercion do they exercise upon it? What subtle impress do they leave behind? Is it possible to make the plants write down their own autographs, and thus reveal their hidden history?[1]

So writes Jagadish Chandra Bose, the scientist who proved that plants were living beings and who sought to create conditions for plants to write their own script (*torulipi*). What Bose seeks is not "autograph" alone; through it, he is seeking autobiography, the "hidden history," the "unsaid" (*abyakta*) of the plants. Only in this plant script will the history of plants be revealed to us, in their "own handwriting." What he is trying to create, therefore, is language.

> If a tree responds to an external injury, how long will it take for this response to materialize? Does the duration of that feeling change in a different situation? Will it be possible to get the tree to write about this time? How does the injury reach the tree's inside? Does it have a nervous system? If it does, how is the speed of the nerve's excitement communicated? Do favorable circumstances cause an increment in its speed?

Do adverse circumstances prevent that? Are there similarities between our nervous system and the plant's? Can the changes in speed be recorded by the plant itself? Do plants have muscles like there are in the human heart? In the end, when death comes to a tree, is it possible to record that moment of nirvana? And does the tree respond fiercely to that moment before falling into an everlasting sleep? Only a history of these different moments, captured by different instruments, will give us the true and uninterrupted history of plant life.[2]

To record this script, Bose would design the electro-optic analogue, the shielded lens antenna, resonant recorder, phytograph, plant photograph, automatic photograph, Bubler instrument, plant sphygmograph, and, of course, the crescograph.

To know the real history of plants, we must go to the plants themselves. That history is mysterious and complicated. To retrieve that history, machines will have to record every moment of a plant's life, from birth to death. The script will have to be written by the plant itself, and it must have the plant's autograph. This is because men are often deceived by their motive.[3]

Bose was not alone in searching for a plant script. Rabindranath Tagore, who called Bose his "first friend," tried to understand "phooler bhasha," the language of flowers.[4] In one of the many songs that Rabindranath Tagore wrote about the mind, often addressing the mind directly, his confusion is evident; he is seeking a language of the mind—what the mind has left behind, that it has no peace, that it is forgetful of the path it took so that it cannot return by the same route again. The speaker tries to listen to the river; he hears it shiver like leaves. He thinks that he'll be able to find his mind if only he could understand the "language of flowers." What is this language of flowers, and why does Tagore turn to it as a metaphor to hold the lack of understanding of the ways of the mind? One notices in the groping for an analogy the same interest, curiosity, and bafflement about this language, the syntax of the mind and the flowering world. It is as if the pathways that the mind takes but cannot retrace, that are its "handwriting," as it were, are like the language of the flowers, which, once in bloom, cannot be "erased" or undone.

"Phota"—or blooming—is the foreign language that Tagore wants to understand. It is as if he knows but does not know, as we find in another song. Every day the flowers bloom in your garden; why don't you feed their honey to my bee-like mind?[5] This line, and those that follow, are about the desire to know the logic and language of the universe and its actions: why he (he calls himself "servant") isn't called to God's assembly even though one is held every day, why God doesn't deign to look at his face when he looks at everything else, etc. Tagore wants to understand the language of indifference by contrasting it with the language of favoritism, blessing, and inclusion; all other beings are blessed by your attention, he asks, why not me? He begins his song with the language of flowers, and he ends it similarly: "you fill the flowers with fragrance."

If I had to identify a refrain of thought in these songs about plant life, it would be this: a curiosity about the *language* of plants, an acknowledgment that their unique form and fragrance *are* their language, a desire to understand both their langue and parole, and a recognition of the impossibility of doing so. We find a record of both his frustration in not being able to learn this language and his acknowledgment of the superiority of the language of flowers in a poem entitled "Phool Photano" ("Making Flowers Bloom"). To paraphrase: None of you will be able to make flowers bloom—no matter what you say or how hard you try, whether you lift it up day and night or injure its stalk, you won't be able to make it bloom. You might make it look faded or pale by staring at it constantly; you might tear apart its petals and branches and throw them to dust; even if the noise and chaos you create might force its mouth open, you won't be able to take its color or rid it of its scent. The one who can make the flowers bloom, he only looks at them—as soon as the rays from his eyes touch their stalks, it's almost like a mantra. He can, only he can make flowers bloom. His breath makes the flowers want to fly, the leaves become wings; they want to crawl on air. All the colors that bloom like the soul's eagerness; it is as if they are calling someone with their scent. He can, only he can make flowers bloom.[6]

Tagore is trying to understand a language that he recognizes, beyond the obvious, as not having originated from a human. How does one understand such a language? Tagore approaches it like a child and a linguist: what is it that leads to expression, whether oral or written? Importing that understanding of communication as reaction and response, an intuition he shares with his friend Bose, he, like the scientist, locates the trigger of this language in light. But he also goes beyond that in locating the personality of the intelligence that has created this script. He is seeking author and intention. What kind of language is this—mother tongue or father tongue? From the introduction that he wrote for Dakshinaranjan Mitra Majumdar's *Thakurmar Jhuli*, a collection of grandma's tales from the small towns and villages of Bengal, we know that Tagore was insistent—and persistent—about the need to collect these stories and the vocabulary of care and intimacy that gives blood to them. He saw this as resistance to the "bideshi" (foreign stories) that had found their way into the Bengali child's imagination:

> Is there anything quite as swadeshi in our country as *Thakurmar Jhuli*? But, alas, even this delightful bag of tales was coming to us not so long ago manufactured in the factories of Manchester. These days, English "fairy tales" have begun to become the only recourse for our children. Our own swadeshi Grandmother Company is utterly bankrupt. If you shake its bags, Martin's *Ethics* and Burke's notebooks on the French Revolution may fall out now and again; but whatever happened to the Royal Prince, the Royal Minister's son, to Bangoma and Bangomi, to the seven kings' precious jewel that lies beyond the seven seas and thirteen rivers? Festivity, jatra, song have all gradually dried up like the dead river, and Bengal's villages, where the current of delight used to flow in various tributaries, have become a wasteland of dry sand. Because of this, the hearts of the grown-ups are becoming hardened, selfish, and mean. And thus our children too have been excluded from delight through no sin of theirs. Why are today's bedtimes so silent? In study rooms, the hum you hear by their kerosene-lit table is the traumatic sound of English spelling. Can a child be happy if you remove its mother's breast from its lips and feed it with only tasteless and nourishing food?[7]

It seems to me that he intuitively thought of the language of flowers in a similar way, as having been produced by the soil and light, in the same way that these stories had been produced by the land.

Understanding is the working definition of love, the reason we ask those we love why they misunderstand us. But if we can't understand the language of flowers, are we misunderstanding them? This is not Tagore's question but mine. In a poem, he asks the dear one whether she understands him: "Tumi morey paro na bujhitey?" He then lists those things whose language is difficult to understand: the river, the sky, the elements, the *dhvani* of the night, and, desperate for understanding—which is shorthand for a shared language—he turns to the language of flowers: "If it were only a flower, round, beautiful, small, blooming to the sun, swinging to the breeze, I would have plucked it with care and tucked it into your dark hair." Not vase, not flowerpot, but hair—that would show care, for that is the language of flowers. Who has—I think in an aside as I sit with the poem—ever seen any woman wearing roots or drying branches in her hair?

Like language, whose nature it is to generate response, the blooming of flowers is a sign that one must understand. The natural world of other plants and animals understands this language as something moving toward fruition, but what is the human to make of this besides the obvious scientific truism? Tagore understands this in various ways. "Flowers have blossomed in the forest; is it right that you should still hold on to your disappointment?"[8] The flower speaks from time to time, as in a short poem where the flower is speaking to the fruit. In an apostrophic poem—I mention this to imagine parts of us calling out to each other, like the flower calls out to the fruit—the flower asks the fruit how far away it is. The fruit addresses the flower, quite surprisingly, as "mahashoy," thus identifying it as male. "Why do you need to call me?," scolds the fruit affectionately, "I live inside you endlessly."[9] Even between the flower and fruit, there seems to be some gap in communication, a lack of shared language.

Where there is a flower, there will be plucking. Is that part of its language too, this intuition of pollinator, this other possible end to its life like its transition to fruit? In a four-line poem, Tagore writes: "I came to the forest to pluck flowers; I don't know what was in my mind; This is not the plucking of flowers; I don't know how I feel, tears fill my eyes."[10] How similar it must be to the language of humans then, if it can move one to tears. We also encounter a different intuition: is plucking flowers right? If we import this to an understanding of language, the question becomes, is plucking flowers ungrammatical? Here is Tagore: "I've made a mistake in plucking the flowers for the sake of my love, my bride; how will I bind you in my sweet bond?"[11] This conscientiousness about plucking flowers, about the sense of it being a mistake, even if it is a retrospective awareness, what is its equivalent in language besides perhaps it being ungrammatical to life?

Language is, of course, not a valve; there is as much speaking as there is listening. How do flowers "listen"? Can flowers listen at all? "There's the rose in bloom, but don't go there," says the speaker to a bee in a Tagore song. For the sweetness of the rose might lead to injury caused by its thorns. "Go to the chapa and shefali flowers instead—tell them about your grief, your pain." The bee refuses: "I won't tell them what I haven't ever before; what lies hidden in my soul I'll share only with the rose; if sharing it with the rose burns me, I will burn from the wound caused by the thorn."[12] This is an unexpected line of reasoning, a language of the bee that only the rose will understand or will, at least, understand better—a secret language whose meanings are only for a very limited and private listener. Is such language a code, meant for only specific "readers"? In all these songs and poems, I notice the same impulse to understand the flower's "script" or language, sometimes as the bee, occasionally as other pollinators, often, analogically, as a language system from the human social world, but most often as a language that is outside both our ability to create and to understand. In such a vocabulary is the intimation of the creator.

Flowers are talkative; they have things to say, often to this creator. They are always in the middle of some manner of conversation in Tagore's poetic universe: the flower says that it is blessed to fall on the earth; and then, addressing god, declares that service for god is its home. "I was born in the dust; forgive me for that, for there's no dust in my soul." Tagore seems to imply this as a form of communication—the falling of the flowers, their service to the creator, and then the downcast eyes of the creator and the concomitant shaking of the branches.[13] On a very simplistic level, this might seem like a justification for the use of flowers to worship the Hindu gods, but I take it to be a little deeper than that: the falling of the flowers to the earth, from where they say they began their journey to rest at god's feet, is a syntax that is familiar to the poetic. I mean the chiasmus, its pointillist circularity that is the form of this life, the flower's and its life's. Why flowers must fall to the earth, why that must be part of its language, as if it were the inevitable period in a sentence, how that "looks" in the plant script is a recurring thought in Tagore, as if the falling of flowers *must* imply something more. In this song, he writes with deep empathy, almost becoming a flower, thinking of staying and falling in relation to light and darkness, as if falling wasn't an act to be imagined as being in scare quotes but as an inevitable dialect, like yes or no: what flower has fallen in this deep darkness, its fragrance has spread to the corners of my sleep; it had come alone, mistakenly, in this darkness to worship sunlight. Why must flowers fall to the ground? And why does it set off a trail of emotions in us? In "Ki Phool Jharilo," a song from the "Prem" (Love) section of *Gitabitan*, Tagore's book of songs, the life of the flower is seen as one of devotion and dedication, and its falling, consequently, a thing of melancholy. In its weak body, one almost dying, it had taken infinite courage and unfruitful devotion: "I couldn't see its form with my own eyes; I don't know its name; in the darkness it went away to the family of stars . . . It doesn't know the 'bani,' it doesn't have the words to explain why it had come to the night's closed door." It's a very moving situation—for the flower, a creature indulged by daylight to turn up at night, to be turned down by the night, all because its language or communion is with light, not its opposite.[14] It feels like having arrived in a foreign language.

What causes flowers to fall and what it means in its unknowable language, whether it is a sign of response at all, is a question that he returns to time and again. In the song "Je Phool Jharey" ("Flowers That Fall"), there is a seeming equipoise about this: "the flower that falls, it falls; flowers keep blooming; the breeze carries them away; the earth dissolves them into itself; it gave fragrance; it gave smiles; its game is now ended; it gave love; is that why it's being ignored?"[15] Falling and blooming, these are the two poles between which Tagore situates his plant language. It is not completely different from the parenthesis we use to mark the arrival and departure of a human citizen. That flowers bloom every day is both a thing of wonder and confusion for Tagore—it must be like language, then, which is polished by some speaker or the other every day. "Every day flowers bloom in gardens; why don't you feed your mind with their honey?," he asks in the song "Nitya Tomar Je Phool Phote" ("Every Day Your Flowers Bloom"). He asks the invisible to become like the flower, like the plant, like the *visvakamal* flower that blooms by kissing the feet and then looks up: "Why can't you be like that flower? Why don't you look at me like that? You fill the flower with fragrance, why don't you give this beggarly heart your daily prasad?"[16]

In a song addressed to the *shiuli* flower that gives Bengal's autumn its peculiar scent, he, after recording his wonder about the power of the night, its lack of light that produces this fragrance, asks, "Kon bhashaye chash bidaye?," in which language do you want us to say goodbye?[17]

He, like his friend Bose, wants the flowers to respond, whether it's asking them for their language of preference in which to hear goodbyes or their names. "Hey bideshi phool," he asks a "foreign flower" in a poem, "What is your name?" Smiling, it shook its head: "I understood that it has no use for names; its identity was its smile." For every question it is asked, the "foreign flower" had the same answer: a smile. In the end, Tagore gives in to his obsessive curiosity: "Hey flower from abroad," I whispered into its ear, "What is your language?" Smiling, it shook its head, and the leaves murmured, I said, "I know, I know, it's the bani, the speech, of fragrance; my breath has filled with the language of your breath."[18] How, then, do we read the language of nonflowering? One might read it as silence; Tagore reads it as the end, the end of song, that space and that pause which is not caesura but the awkward and sublime confusion of "The music in my heart I bore / Long after it was heard no more. I've exhausted the flowers; my song has come to an end. God, take my last offering." After all the offerings, of light and lightlessness, there is that familiar language: the flowers smiling and losing their petals, the moonlight smiling and disappearing.[19]

In "Sakhi Bhabana Kaharey Bawley," one of his most well-loved songs about the nature of thought and love, both of which are reliant on language, he tries to distinguish the world of flowers from that of humans, and to do so he turns to emotions: "the elements and the moonlight and the rivers do not know sadness and tears; they only smile, they only sing; the flowers laugh and play and the petals fall as they smile, as they die."[20] Song should bloom like flowers.[21] In that expectation, almost a season in Tagore's world, one is forced to think whether, by extension, the flower script could be like music notation.

Tiny flowers fascinate him: "I only make garlands of the smallest flowers; they are quick to die—if that be the case, let it be; I will remember them till eternity. If this garland helps those who live in the darkness to find a moment of happiness, if they can forget the pain of attachment temporarily . . . These tiny flowers, they bring freedom with their fragrance—if by looking at these tiny flowers one is reminded of the boundless world and the boundless sky." In these flowers, in their smallness, he sees the signature of the world, its language. In these flowers, he finds the language of the universe.[22]

He is seeking history, geography, and geology in this flower linguistics. In a poem written for children, he sets out to give us, as the title makes clear, "phooler itihash," the history of flowers. Flowers speak in them, in a language that they share, the flower and the honey collector. The honey collector asks for honey; the flower gives everything it has; the wind asks for fragrance; the flower gives it all; the honey collector returns; the flower, now wilting, has nothing left to give; the wind returns; the flower says, "What more do I have . . ." Tagore's implication here is not just moral; it is linguistic. The human and the wind are monolingual; they speak in a language that the flower has to understand, but they have not bothered to understand the flower's language; they cannot see that the flower has exhausted itself through its generosity, that it is wilting, that it has nothing left in it to give anymore, not even to itself.[23] Its death is the loss of language. This understanding of an inequitable understanding of languages permeates several other poems: "Chhi chhi what have you done, my friend, the kamini buds were lighting up the forest; scared of the human touch, trembling, their corpses now lie on the forest floor. You know that the kamini flower is a virgin; one has to watch it from a distance; it shouldn't be touched; it emits a sweet scent; it cannot bear the intimacy of human breath. The human does not—cannot—understand the language of flowers, their codes, their semiotic, their index, their lack of consent."[24]

Sometimes Tagore imagines the language of the lover as similar to the language of plants: "'Tumi kon kananer phool,' which forest are you a flower of, in which sky are you a star? I've seen you somewhere, as if in a dream; I've forgotten when you left, when you looked at me. You don't speak; you look and go away; in the moonlight, you smile and melt away . . . Notice how the woman, silent, who will not speak, is unconsciously being imagined as a flower—you don't speak; you smile and melt away."[25]

It is a compulsive tic in writers to hear things in quotation marks. When I read Tagore, I sometimes wonder whether he saw the blooming of flowers as "direct speech," flowers as beings that must exist inside quotes. He, of course, doesn't say this anywhere—these are my slightly wayward thoughts—but his permanent sense of wonder about who is responsible for the blooming of flowers leads to the inevitable thought: "'Aeto phool ke photalo kananey?' Who made all these flowers bloom in the forest? Whose language is this, these flowers in bloom, what does it mean?"[26]

That the flowers are a response to or in communication with the seasons annotates his understanding of this language, so that it seems that this is a dialogue, perhaps even slightly Socratic in nature: "'Badol diner pratham kadam phool korechhey dan / Aami ditey eshechhi Sraban-er gaan . . .' The first kadam flowers of the season are a gift; I've come to offer the songs of Sraban"—this serve-and-volley between the season and the flowers is endless and infinite. The human, feeling deprived of this language, wants to change the dialogue into a discussion—hence his intervention, through the "Sraban-er gaan," singing the songs of the rainy month of Sraban. And then the parallels between the response of the flowers and the human: the rains cause an outpouring of song in him as they do on the branches filled with flowers. Flowers as song, as sur, musical notes; another language, but one that can do without a script.[27] But flowers are not just a response, they are triggers too: "The spring flowers make me 'April-mad,'" he says in a poem. "My gray hairs have turned black, my brains are stuck, everyone laughs at my condition."[28]

If flowers—and flowering—is indeed a language, what does one do with such a language? "Spring has strung a garland of flowers for me, to celebrate my victory," writes Tagore.[29] If the garland is a resting place for flowers that are now detached from their parent, is the garland like a sentence where words have come together in a sequence? Is it this collective intuition that is behind the phrase "garland of words"?

Unlike Tagore, I went to the leaves. To identify every kind (type) of leaf as a letter of the alphabet requires a willed imagination. Not ever having had it, I do not miss it. I had something else, more a lack than a virtue actually—a fascination for language, often for etymology. Where had the word "script" come from? From "scribere," meaning "to write." Who had ever looked at a leaf and imagined that it could write? One knew, from both history and intuition, that the leaf had, in fact, provided a space to write on, the reason perhaps why the word for leaf and page is the same in Bangla: "pata."

I think at first it was the heart. Hearts for leaves, tiny, tiny hearts crawling up a stem. Sometimes it seems that they mimic the raindrops. They remain close to the earth, like little people, unambitious for height, distrustful of the vertical. They will take over any place that is given to them—they are like the heart, which can accommodate everything and everyone. Hearts whose chambers we cannot see, hearts connected by string-like stems, so many of them, light green when young, maturing into a darker color when older, the opposite of human hair, which goes lighter as we age. When hung in pots, they pour out of it in abundance, so that from a distance, they might resemble a leaking pot with water sprinting out of it. I collected them when they dried or when birds tore the stems and left them behind.

Since it was the shadows of the tiny hearts on the walls that had at first seemed like a script—a script whose meaning we were still trying to decipher—it was to them I went. Coating their dead branches and leaves with black watercolor paint, I pressed them onto white paper. But I'm not the sun. My pressure is not weightless like light; if a plant is exposed to sunlight unequally, it will reveal itself in the differences in shape and color. But even as I tried my best to distribute pressure equally on the paint-coated plant, the results were disastrous. I mean, they were wrong. The "script" was more a record of my body and mind than it was a record of the plant. If the plant had wanted to "write," it would have. Why was I, like a parent or schoolteacher, forcing the script on a species—on a world—that had neither any urge nor any use for it?

But I was stubborn. And I was restless. Having no training in making art, I had no idea about waiting time. I moved the books under which I'd placed the paint-soaked plant. What had I expected? Why had I expected to see something resembling the shadow of these plants? What I had instead was something like a sea beach full of footprints, or an ant trail or crows on an electric wire. It is only now that I see that my first spontaneous comparisons were with animals. The leaves had had no desire to "write"—I was only trying to record their form the way an ink-smeared fingerprint is a record of my finger. Was I my finger? Was the page filled with my fingerprints my personal alphabet of ten letters? Why, then, was I taking the form of the leaves to be a script? How was one to read it? Left to right, right to left, top to bottom, bottom to top? As I tried to record or imagine the plant script, I began to notice the arbitrariness of our language systems. Who had decided on these orders of looking and observing? Should I look at a tree from its roots to its top or top down? There was another reason for the confusion: the impress of the leaves both looked and did not look like a script. By "script," I am thinking of something that has emerged from the human.

There was also something else. The patterns on the piece of paper were—I find it hard to say it—actually analogous to a death mask. In other words, this was a death certificate. The leaves had been dead, their movements a thing of the past—this wasn't a record of response, which is how I understand language, a serve-and-volley between our circumstances and ourselves, moving, as they often do, between the registers of "aha" and "shit." My experiments were childish, or, if I must be generous to myself, a bit like a child trying the letters of an unfamiliar alphabet, letting the leaves and branches dry, picking them up from places, coloring them, pressing them on paper. It did not take me very long to realize, or to come to accept, that mine was like a death script. I tried not to blame myself. For I had tried to let living leaves and branches make their impress on white paper. Apart from the fact that it felt like I was asphyxiating them by dipping them in color, even if it was watercolor, there was the problem of dimension—it was impossible to reduce living forms to the flatness of paper. This made me wonder about the idea of script itself: how can a script, any script, hold my being in it?

Language—and script—are living things; when they are not alive, we find the cliché, dead language. Was this a plant cliché then? Even if it was, it wasn't the plant's cliché. This immediately took me to thoughts about rhetoric, its significance in human languages, and whether anything comparable existed in the plant world. My first stop was at "alankara"—ornamentation, its persuasions and dissuasions. Poetry, at least classical poetry, seemed to be unimaginable without it—sometimes it seemed a bit too much, distracting, even an indulgence. Surely flowers were not ornamentation for plants but a necessity, even though humans had used them as ornaments and decoration. Carrying my understanding of rhetoric and prosody to the plant world did not help me to understand their "language" better, but traveling in the opposite direction, importing this understanding of nothing being superfluous or unnecessary in the plant world to my understanding of language, particularly poetic language, helped me to discover things about the latter that I might not have otherwise.

I had to suppress my proclivities—no, the response of plant life to spring wasn't the equivalent of an exclamation mark, neither was a fallow period of nonflowering a period. Punctuations derived far too much from the human body, its reflexes and reserve. The anatomy of plants was different. Language—all kinds of human language, the script, drawing, music—was inevitably and compulsively synecdoche. It is impossible to think of plant functioning in terms of the grammar of the synecdoche, the part for the whole. I had tried with living plants and branches. There's a disobedience in the form that doesn't commit to two-dimensionality, which is, after all, what the "script" is. That is why shadows are not our reflections. I became, by Plato's indictment, twice removed from the truth when I started taking photos of the shadows. I have no reasons to offer—why did I think of plant shadows as script when I never imagined my own shadow or the shadows of animals as our script? I feel embarrassed. What was I trying to understand as form? The anatomy? I had recorded the sounds of branches and leaves in response to the wind, but it excluded large tree trunks that did not make any sound audible to my ears. Wasn't that part of its language too? It would be wrong to call it "music," as it is to call anything produced by trees "art."

What was the plant script and what did it mean, this urge to imagine and ascribe literacy to plants? The shadows of branches triggered mental games with oneself—how they resembled certain letters of the alphabet: A, W, M, N, H, K, L, T, V, and, of course, Y. There were no branches or leaves or flowers that looked like G or Q—I couldn't help noticing this and aligning it with the first words that came to me: "God" and "Queer." No B, P, or D either, I realized soon after. My thoughts did not have any meaning—no consequence, no trajectory, nothing. But they were not idle either. What did it mean to force similarities of an alphabet between two species, one much older than the other?

It was silly, and yet it was impossible to deny this urge to find a shared grammar of existence, for what else is language? I noticed the way rain fell on plants and trees—there was a script there, as temporary as a child's slate, vanishing and changing quickly, but a script nonetheless. I tried to check whether the raindrops fell similarly on humans—humans and plants get wet unequally, I had to pause the thought. There were others who were looking for other plant scripts to read—scientists were now reading plant rings to study the history of pollution, just as Sylvia Plath had found the history of winters soaked and seeped into this script:

> The wet dawn inks are doing their blue dissolve.
> On their blotter of fog the trees
> Seem a botanical drawing—
> Memories growing, ring on ring,
> A series of weddings.[30]

Have trees ever been ungrammatical?

Notes

1 Jagadish Chandra Bose, *Abyakta* (n.p.: Createspace, 1922).
2 Bose, *Abyakta* (translation by the author).
3 Bose, *Abyakta* (translation by the author).
4 Rabindranath Tagore, "Tui phele esechhis kare," 1915, https://www.tagoreweb.in/Songs/prem-235/tui-phele-esechhis-kare-5684.
5 Rabindranath Tagore, "Nutya tomar je phul phote," 1913, https://www.tagoreweb.in/Songs/pooja-233/nutya-tomar-je-phul-phote-4745.
6 Rabindranath Tagore, "Phul photino," n.d., https://www.tagoreweb.in/Verses/kheya-54/phul-photano-1839.
7 Rabindranath Tagore and Dakshinaranjan Mitra Majumdar, ed., *Thakurmar Jhuli* (1907; n.p.: n.p., 2012).
8 Rabindranath Tagore, "Bane eman phul," 1883, https://www.tagoreweb.in/Songs/prem-235/bane-eman-phul-5749.
9 Rabindranath Tagore, "Phul o phol," n.d., https://www.tagoreweb.in/Verses/kanika-31/phul-o-phol-1838.
10 Rabindranath Tagore, "Ami phul tulite," 1933, https://www.tagoreweb.in/Songs/prem-235/ami-phul-tulite-5721.
11 Rabindranath Tagore, "Phul tulite bhul karechhi," 1929, https://www.tagoreweb.in/Songs/prem-235/phul-tulite-bhul-karechhi-5470.
12 Rabindranath Tagore, "Golap phul phutiye ache," 1878, https://www.tagoreweb.in/Songs/prem-o-prakriti-258/golap-phul-phutiye-achhe-6378.
13 Rabindranath Tagore, "Phul bale dhanya ami," 1933, https://www.tagoreweb.in/Songs/chadalika-gitabitan-514/phul-bale-dhanya-ami-9616.
14 Rabindranath Tagore, "Ki phul jharila," 1925, https://www.tagoreweb.in/Songs/prem-235/ki-phul-jharila-5655.
15 Rabindranath Tagore, "Je phul jhare," 1885, https://www.tagoreweb.in/Songs/prem-235/je-phul-jhare-5764.
16 Rabindranath Tagore, "Nitya tomar je phul phote," n.d., https://www.tagoreweb.in/Songs/pooja-233/nutya-tomar-je-phul-phote-4745.
17 Rabindranath Tagore, "Shiuli phul shiuli," n.d., https://www.tagoreweb.in/Songs/prakriti-236/shiuli-phul-shiuli-5239.
18 Rabindranath Tagore, "Bideshi phul," n.d., https://www.tagoreweb.in/Verses/purabi-77/bideshi-phul-1386.
19 Rabindranath Tagore, "Phul to amar phuriye gechhe," n.d., https://www.tagoreweb.in/Verses/gitali-64/phul-to-amar-phuriye-gechhe-6157.
20 Rabindranath Tagore, "Sokhi bhabna kahare bole," 1881, https://www.tagoreweb.in/Songs/natyogiti-516/sokhi-bhabna-kahare-bole-10111.
21 Rabindranath Tagore, "Phuler moto apni phutao gan," https://www.tagoreweb.in/Verses/gitanjali-62/phuler-moto-apni-phutao-gan-3175.
22 Rabindranath Tagore, "Chhoto phul," n.d., https://www.tagoreweb.in/Verses/kori-o-komol-5/chhoto-phul-2200.
23 Rabindranath Tagore, "Phuler itihas," n.d., https://www.tagoreweb.in/Verses/shishu-52/phuler-itihas-1841.
24 Rabindranath Tagore, "Kamini phul," n.d., https://tagoreweb.in/Others/shaishab-sangit-172/kamini-phul-262.
25 Rabindranath Tagore, "Tumi kon kananer," 1886, https://www.tagoreweb.in/Songs/prem-235/tumi-kon-kananer-5740.
26 Rabindranath Tagore, "Eta phul ke photale kanane," 1883, https://www.tagoreweb.in/Songs/natyogiti-516/eta-phul-ke-photale-kanane-10139.
27 Rabindranath Tagore, "Badal diner pratham kadam phul," n.d., https://www.tagoreweb.in/Songs/prakriti-236/badal-diner-pratham-kadam-phul-5218.
28 Rabindranath Tagore, "Epriler phul," n.d., https://tagoreweb.in/Verses/prohasini-127/epriler-phul-5783.
29 Rabindranath Tagore, "Basante phul gathlo," 1915, https://www.tagoreweb.in/Songs/prakriti-236/basante-phul-gathlo-5301.
30 Sylvia Plath, *Winter Trees: Poems* (1971; New York: Harper Collins, 2016).

12

A HYPHEN BETWEEN TREES

Derek Walcott's Canoes

HANNAH RACHEL COLE

> The canoe is a hyphen between centuries
> between generations, between trees.[1]

> Tree! You can be a canoe! Or else you cannot![2]

In "100 Years Ago (Carrera)," the Saint Lucian Nobel laureate Derek Walcott describes a canoe in the Gulf of Paria as "a hyphen between centuries / between generations, between trees."[3] This formulation, which appears in his 2016 collaboration with Peter Doig, *Morning, Paramin*, imagines the red wooden vessel that spans the width of Doig's painting as a grammatical mark running across historical time and ecological space (Figure 12.1). These two lines form the sole rhyming couplet at the conclusion of what is otherwise a prose poem (notably the only such composition in the collection), and they are set off from the rest of the poem by an em dash, which resembles and extends the form of the hyphen. The poem invites the question: why the sudden shift to verse? And why the closing reference to trees in response to a painting where vegetation appears only indistinctly in the background? At stake is Walcott's larger investment in the power of the canoe as a mediating device between the histories, geographies, and arboreal ecologies of the insular Caribbean. The switch to verse in a poem that otherwise eschews it suggests a visual and symbolic correspondence between the canoe, the hyphen, and the tree whose hollowed trunk becomes the vessel's raw material. Indeed, the final clause in Walcott's series, "between trees," speaks to his long-standing attention to the internal dynamics of Caribbean forests across his oeuvre. The transition to meter in this closing couplet evokes the rhythms of relation between the arboreal species of the tropical highlands, whose terrestrial life is "hyphenated" with the aquatic

FIGURE 12.1
Peter Doig, *100 Years Ago (Carrera)*, 2001. Musée national d'art moderne / Centre de création industrielle © Art Resource, New York.

ecology of the Caribbean Sea by the process of canoe construction.

This chapter takes its cue from Walcott's conception of the hyphen "between trees" to consider the ecological function of the dugout canoes in his 1990 Hellenic epic *Omeros*. It argues that the construction of the vessels at the poem's incipit should be read as a hyphenation of the diverse tree species of the Saint Lucian forest as well as between the conditions of multispecies thriving and useful matter. This reading follows Maria Cristina Fumagalli in her understanding of Walcott's canoes as references to Amerindian seafaring practices across the insular Caribbean.[4] As such, they hyphenate the postcolonial present of late twentieth-century Saint Lucia with the precolonial legacy of boatbuilding across the archipelago and, Walcott suggests, the intervening history of the Middle Passage. Moreover, the *gommiers* mediate between touristic and local understandings of Saint Lucia as well as between the island's terrestrial geographies and their aquatic counterparts.

In reading "between trees," this chapter follows George B. Handley's claim that Walcott "consistently searches for the merger between human and natural history."[5] This union materializes in the hyphenating function of the canoe. The reading's plant humanities approach builds upon Elaine Savory's study of Walcott's flora, which indexes imperial history and environmental precarity at once.[6] As Elizabeth M. DeLoughrey, Renée K. Gosson, and George B. Handley have argued, "in the battle against amnesia induced by colonialism's erasures, the deterritorialization and transplantation of peoples, and even natural disasters, the Caribbean writer often seeks nature as an ally."[7] The canoe proves a particular ally to Walcott in his effort to invoke Indigenous presences in twentieth-century Saint Lucia against the risk of erasure, while the violent story of the canoes' creation speaks to the ongoing dangers of biodiversity loss and extinction that threaten the island's forests.

While this chapter occasionally refers to Latin binomial nomenclature for the sake of clarity, it does so in recognition of botany's imperial history in the region. Indeed, Walcott argued in the 1973 essay "Isla Incognita" "that our only true apprehensions are through metaphor, that the old botanical names, the old processes cannot work for us."[8] Instead of assuming the "arrogance of an Old World botanist," one must learn each plant's "creole or country name according to

FIGURE 12.2
Taíno dugout canoe, from Girolamo Benzoni, *La historia del mondo nuovo*, 1565, woodcut. Library of Congress, Washington, D.C.

its properties."[9] In other words, plants should be understood through local knowledges that attend to their practical uses and figurative meanings instead of the imposed language of taxonomy. The reading of Walcott's canoes as hyphens "between trees" adopts this approach to tease out the metaphorical associations between species, yet taxonomy is indispensable for translating his references to a potentially unfamiliar audience.

With that said, it is worth noting that Walcott refers variously to the vessels in *Omeros* as canoes, pirogues, and *gommiers*. "Canoe" and "pirogue" are likely loanwords of Kalinago and Taíno origin, while *gommier* derives from the French, meaning "gum tree."[10] The common name *gommier* can denote either of two West Indian tree species, *Bursera simaruba* (*gonmyé modi*) or *Dacryodes excelsa* (*gonmyé*), the latter of which is traditionally used to make dugout canoes (Figure 12.2).[11] Its "gum" is the flammable resin that exudes from the bark of the tree, which can be burned as incense or used to start fires in the rainforest.[12] In Saint Lucia, the word *gommier* can refer to either the tree or to the boat constructed from its wood. The fact that all three terms (canoe, pirogue, and *gommier*) appear liberally throughout *Omeros* speaks not only to the poetic necessity of an extensive lexicon within the *terza rima* scheme of the epic poem but also to the function of the wooden dugout vessel as a line of connection, a hyphen, between Amerindian, European, and Afro-Caribbean cultures.

Amerindians have used dugout canoes for travel and fishing since the Ceramic Age, but their selection of trees has shifted over the centuries.[13] While Christopher Columbus, Raymond Breton, and others report seeing canoes constructed from red cedar (*Cedrela odorata*, known locally as *acajou*), *gommier* has since supplanted cedar as the material of choice, perhaps as a consequence of the overharvesting of the latter.[14] Today, the Kalinago of Dominica are one of the only remaining Amerindian groups that continue to manufacture traditional canoes, and while the *gommier*'s use for commercial fishing is declining in the Lesser Antilles, it continues to feature in annual sailing races between Saint Lucia and the neighboring French island of Martinique.[15] For the Martinican philosopher Édouard Glissant, these "glorious regattas or traditional races of *yoles* and *gommiers*" function as a counterpoint to the dark history of the Middle Passage.[16] His *Poetics of Relation*, published the same year as Walcott's

FIGURE 12.3 Jean Baptiste Du Tertre, *Histoire générale des Antilles habitées par les Français*, 1667.

Omeros, suggests that the splendor of the *gommier* regattas contrasts with the abyssal memory of the Atlantic slave trade. At the same time, the very fact that the canoes are "navigating the green splendor of the sea" still conjures "these lowest depths" of history.[17] The *gommier* thus registers multiple connections in the Lesser Antilles: to the Amerindian seafaring knowledges that have survived erasure; to the horrors of the Middle Passage that left untold numbers of Africans drowned at sea; to the vanishing lifestyles of traditional Afro-Caribbean fishermen; and to the changing composition of the forests where the trees are harvested (Figure 12.3).

What might it mean to conceive of a canoe as a hyphen between trees? The word "hyphen" is derived from the Greek *huphen*, meaning "to tie together." Yet, as Pardis Mahdavi has shown, hyphens can divide as much as they connect.[18] They can entreat the reader to interpret two words as a single entity, or they can separate a word's syllables across a line break. This dual function is at play in the mediatory function of the canoe, which links the diverse arboreal ecosystem of the Saint Lucian forest vis-à-vis the associative logic of the poem. Yet the *gommier* also marks a break between the living trees and the vessels they become. Traditionally, their trunks are converted into seaworthy crafts by a multistage process in which they are felled in the highlands, hollowed out, transported to the coast, reshaped with fire and stones, and fitted with ribs, benches, and paddles.[19] This procedure is also a ritual ontological metamorphosis. While there are significant gaps in anthropological understandings of the symbolic significance of canoe construction for precolonial Amerindian boatbuilders, evidence points to feasting, drinking, and baptism ceremonies as ways of anchoring the trees' transformation in the spirit world.[20] Walcott's interpretation of this series of rituals in *Omeros* conveys its duality: boatbuilding connects the trees to the cosmos, to each other, to the fishermen, and to the sea, yet it also divides the felled organisms from the soil and the forest ecosystem that produced them.

This process is the focus of the invocation of Book One of *Omeros*, in which the fisherman Philoctete recounts the history of how the canoes were built. In the first lines of the epic, this story facilitates the poem's mediation between tourists and locals: "'This is how, one sunrise, we cut down them canoes.' / Philoctete smiles for the tourists, who try taking / His soul with their cameras."[21] The tourists take the place of the Muse in the form of the classical epic's invocation.[22] While they never appear within the poem itself, they lurk at its edges, and their interest in the history of the vessels is to add a little local color to an otherwise interchangeable vacation.[23] The picturesque sight of the long dugout canoes is destined to be captured, it is suggested, in the lenses of their cameras. Yet while Philoctete's explication of the crafts' construction is directed at the vacationing crowd, it links them to a locally rooted West Indian history and cosmology that exceeds their frame of reference. His account of the pirogues is thus hybridly touristic and local, a hyphenated history that at once connects and divides the Saint Lucian storyteller from his foreign audience.

At the same time, Philoctete's language interweaves geographies of land and sea. The phrase "cut down them canoes" reverses the relationship between tree and watercraft, suggesting that the *gommiers* were always already canoes, even when their trunks were rooted in the soil. This fluid negotiation of terrestrial and aquatic geographies operates "tidalectically," in the terminology of Barbadian poet Kamau Brathwaite.[24] Indeed, Walcott once described canoes in an interview as "something organic, because that tree is like wood and water; it is made out of wood and is now afloat and moving on the sea."[25] In this sense, he considers the vessel to be "two elements" brought together through the process of its construction.[26] The poem's trees, figured as canoes rooted on the land, epitomize this hybridity, connoting a fluctuating interface of land and sea.

Walcott's representation of the felling of the trees intertwines the multiple species of the Saint Lucian forest. Philoctete continues his account for the tourists: "Once wind bring the news / to the *laurier-cannelles,* their leaves start shaking / the minute the axe of sunlight hit the cedars, / because they could see the axes in our own eyes."[27] Surprisingly, this account begins with the *laurier-cannelle,* not the *gommier* tree that is generally used to build canoes. Mac Fenwick has conflated the two species, defining the *laurier-cannelle* as "a native West Indian species of *gommier* (*Dacryodes excelsa*), which was revered by the Arawaks and is now extinct through much of the West Indies (including on Walcott's home island of St Lucia)."[28] In fact, the tree known as *lowyé kannèl* in Creole is actually *Aniba ramageana,* and while it is considered "vulnerable" by the International Union for Conservation of Nature, it remains extant in Saint Lucia.[29] It was indeed overexploited as wood for furniture-making, but its population is currently recovering from the brink of extinction.[30] There is no evidence in the literature on Amerindian canoe construction to suggest that *laurier-cannelle* has ever been used as a source material for boats in the region.[31]

While this may seem a minor distinction, it points to the ecological function of Walcott's watercraft. Something more complex than "an act of ecocidal violence" is happening in the poem's opening, although the "murder" of the trees to build canoes is indeed violent.[32] The biodiverse arboreal ecosystem of the highland forest is mediated by the construction of the canoes, which at once entwines the trees associatively and separates them from the soil and each other. In fact, the *laurier-cannelle* is not the only species that substitutes for *gommier* in the text; cedars also figure interchangeably. In the excerpt above, the wind communicates the news of death to the *laurier-cannelles,* which begin to shake as the cedars respond to the axe-like shaft of sunlight. This set of responses to stimulus is sympathetic, suggesting the possibility of communication and even premonition to be shared between trees.[33] For them to perceive the axes "in our own eyes" suggests a model of nature in which humans are not the only agents of consciousness. Instead, perception is distributed among the denizens of the forest, and the ferns become complicit in the killing: "Wind lift the ferns. They sound like the sea that feed us / fishermen all our life, and the ferns nodded 'Yes, / the trees have to die.'"[34] Once more, the sea is figured on the land through the oceanic sound of the ferns, implying a tidalectic relationship between the two geographies. For the ferns to consent to the death of the trees suggests something other than a utopian relationship among the varied flora of the mountain, gesturing at the interspecific competition that structures the natural world. It also implies that Walcott's nature is not a romantic Eden; to the contrary, it can be the site of conflict and pain.[35]

Sneharika Roy has noted the slippage in Walcott's species in this passage, contending that "the discrepant association between the cedar trees and the gommier can be explained by the fact that the incipit is filtered through a narrative consciousness that incessantly oscillates between past and present, between epic memory and Caribbean postcolonial present. . . . The link between the cedar and the gommier is therefore associational and imaginative."[36] I follow Roy in recognizing a paradigmatic relationship between the species that operates across the centuries. Indeed, the initial reference to the *laurier-cannelle*, which was harvested to near-extinction, resonates with the extinction threat to the traditional practice of Saint Lucian boatbuilding that will appear later in the poem, strengthening the logic of association between the two species despite the fact that *laurier-cannelle* is not used for canoe construction.[37] Yet while Roy emphasizes the temporal distinction between the trees, it is worth noting that they exist synchronously within modern Saint Lucian woodlands. While *Cedrela odorata* is no longer the material of choice for canoes, it remains a persistent, if threatened, feature of the island's semievergreen forests.[38] Therefore, the link Walcott is drawing is synchronic as well as diachronic, hyphenating contemporaneous species in ecological relation.

Walcott has a capacious notion of the trees' entanglement. The sawyers must forcibly rip off "the net of vines" that binds the felled trees "to this earth," suggesting their enmeshment within a multispecies plant community.[39] Movingly, he evokes the common language of the trees, which is lost through their uprooting:

> The bearded elders endured the decimation
> of their tribe without uttering a syllable
> of that language they had uttered as one nation,
>
> the speech taught their saplings: from the
> towering babble
> of the cedar to green vowels of *bois-campêche*.
> The *bois-flot* held its tongue with the *laurier-*
> *cannelle*
>
> the red-skinned logwood endured the thorns
> in its flesh,
> while the Aruacs' patois crackled in the smell
> of a resinous bonfire that turned the leaves
> brown
>
> with curling tongues, then ash, and their
> language was lost.[40]

Here, the "decimation" of the trees to build canoes is explicitly linked to the murder of Indigenous peoples across the Caribbean. In this sense, the loss of arboreal language produces an elegy for the loss of Amerindian cultures and languages. Walcott indicates that the different species of trees (cedars; *bois-campêche*, also known as logwood; *bois-flot*; and *laurier-cannelle*) once spoke a common language, inculcated to their kin as saplings and articulated through their trunks and foliage as "towering babble" and "green vowels."[41] Even the fire becomes a site of enunciation, its crackling consuming the organic matter of the forests while communicating in its noise the "Aruacs' patois." A patois implies that the lost language is a hybrid, yet the languages at play are unclear. While suggesting at once the fusion of Arawak language with its British, French, and African counterparts and the French-based Creole spoken widely in Saint Lucia, this reference also implies a vernacular at the nexus of the multiple vegetal species of the forest. Moreover, the "resinous" quality of the fire speaks to the viscous exudations of trees like cedars and *gommiers*, revealing how the matter of the varied species is amalgamated in the acts of felling and burning, which ironically fuse them together as they are divided from the earth. The language of the forest is articulated in the moment of its own destruction, which transforms the leaves of differing trees into an undifferentiable heap of ash.

As the trees are hyphenated to each other through the building of canoes, their metamorphosis is also linked to a different kind of craft: that of Walcott's poetry. In this sense, the mediatory quality of the canoes as hyphens between trees also generates a commentary on the transformation of Saint Lucian nature into Walcott's own literary vessel. Savory notes in her reading of *Tiepolo's Hound* that "[l]eaves have a metapoetic quality as in the leaves of a book, so their prevalence helps to connect Walcott's poetics and the life of the environment which inspires it."[42] Here, the burning of the leaves of different trees suggests their rebirth in the epic poem itself, which makes a point of enumerating their local names so that they survive their own destruction. Symbolically, the felled trees are transformed into the pages that the reader holds in her hands, and their difference is inscribed within the language of the text, even as the uniform matter of wood pulp dissolves their diversity.

At the same time, the felling of the trees is associated with the loss of the ancient gods. Walcott writes: "These were their first pillars that fell, leaving a blue space / for a single God where the old gods stood before / The first god was a gommier."[43] In this passage, the clearing of the trees stands in for the replacement of the Amerindian and African pantheons with monotheistic Christianity. This transition is linked implicitly to the loss of biodiversity, suggesting that monotheism is connected to the simplification of forest ecosystems. Yet why would the first god be a *gommier*? Beyond the alliterative flourish, the poem positions the *gommier* at the origin of the universe as well as at the origin of the poem. The suggestion is that the death of the tree is also the beginning of a new world signaled by the opening of the epic, so that its felling forms a hyphen between the two. The fact that the poem is made possible by the removal of the trees recalls the histories of genocide and enslavement that initiated the formation of creolized Caribbean cultures. As Walcott has noted elsewhere, "geographically and historically, the Caribbean origin is of slavery and indenture."[44] In *Omeros*, he writes into the space formed out of these deadly origins, positioning the postcolonial epic in the wake of transformative violence in both human and vegetal terms.

The scene of the *gommier*'s death is framed as a ritual sacrifice that enmeshes cedars and *gommiers*, sea and land: "Now, over the pastures / of bananas, the island lifted its horns. Sunrise / trickled down its valleys, blood splashed on the cedars, / and the grove flooded with the light of sacrifice. / A gommier was cracking. Its leaves an enormous / tarpaulin with the ridgepole gone."[45] The horns of the island are its Pitons, the two volcanic spires that mark Saint Lucia's southwestern coast. As day breaks, "blood splashed on the cedars," yet it remains unclear whose blood is being shed. Is this the resin of the *gommier*, or is the bright coloration of the sunrise producing a blood-red color in concert with the sacrifice taking place in the grove? Regardless, the cedars are marked by the death of the *gommier* trees, emphasizing the intimate relation between the separate species. The image of the leaves as a tarpaulin without its ridgepole suggests the loss of the tree's shade, which is rendered inoperative by the absence of its trunk. The implication is of the shady tree's place within an intricately balanced ecosystem where it might shelter other species. At the same time, it figures the form of a sail, suggesting the new form the *gommier* will take. In the subsequent tercet, this analogy becomes increasingly explicit as Walcott refers to the trunk as an "angling mast."[46] Its oceanic future becomes evident in the "waves of huge ferns . . . nodding to the sea's sound" and "the ground shudder[ing] under the feet in waves."[47] The liquid image of the splashing blood is linked to the watery fate of the felled *gommier*, and its sacrificial death makes tangible the already potent intermingling of sea and land.

While the maritime future of the tree is foreshadowed, its suitability for life at sea remains uncertain. Achille, struggling to sever it, prays, "Tree! You can be a canoe! Or else you cannot!"[48] This pun suggests the minute difference between the success or failure of Achille's endeavor: the distinction between the tree remaining in the ground and becoming a canoe is a matter of two letters. Moreover, the written form of the word subtly alludes to another local species, *Bois canot*.[49] The suggestion in this orthographic wordplay is that the arboreal species in the sylvan ecosystem are defined in negative relation to one another, just as the dugout canoe is defined in negative relation to the raw material of the tree from which it is constructed. Like words connected by hyphens, the trees are situated in the linguistic tension between unity and separation, and their connection is mediated by the incipient craft.

Achille's project of felling prevails, and the trees become amenable to their new purpose at sea: "The logs gathered that thirst / for the sea which their own vined bodies were born with. / Now the trunks in eagerness to become canoes / ploughed the breakers of bushes, making raw holes / of boulders, feeling not death inside them, but use— / to roof the sea, to be hulls."[50] The tone shifts abruptly from mournful elegy for the dying forest and the memory of its Amerindian inhabitants to jaunty enthusiasm for the seafaring to come, suggesting that the ancient trees have become successfully conscripted into the nautical ambitions of their human loggers. The implication in "thirst," "eagerness," and "feeling" is that the felled trees are animated by the chance to prove themselves useful at sea, recalling Walcott's claim in "Isla Incognita" that "without use [a plant] remains anonymous."[51] In contrast, through their metamorphosis into canoes, the trees acquire new names and purposes. Indeed, Achille's craft is christened "*In God We Troust*" in reference to the celestial past of the *gommier*. Ultimately, the trees collectively "agreed with the waves to forget their lives as trees."[52] The effect of this transformation, which positions Hector and Achille as dueling protagonists of the epic mounted on their crafts, is to emphasize the mutability of forest life as always subject to becoming otherwise. The species of the trees become irrelevant as they are transformed into wood, and while Walcott will continue to refer to the boats as "*gommiers*," we do not definitively know the taxon of their raw material. Yet if the canoes agree to forget their pasts as trees, Walcott's incipit ensures that the reader does not, even as the vessels become defined, like all "true apprehensions," "through metaphor."[53]

As the poem progresses, the canoes will traverse not only ecosystems but continents and centuries. *In God We Troust* ferries Achille on an oneiric journey to visit his ancestors in a village in tropical Africa. After the village is attacked in a slaving raid, he encounters the local griot, whose body takes on the form of a canoe:

> The griot crouched there. Warm ashes made his skull white
> over eyes sore as embers, over a skin charred as coal,
> the core of his toothless mouth, groaning to the firelight,
>
> was like a felled cedar's whose sorrow surrounds its bole.
> One hand clawed the pile of ashes, the other fist thudded on
> the drum of his chest, the ribs were like a caved-in canoe
>
> that rots for years under the changing leaves of an almond
> while the boys who played war in it become grown men who
> work, marry, and die, until their own sons in turn
>
> rock the rotted hulk, or race in it, pretending to row
> as Achille had done in the manchineel grove as a boy.[54]

Here, the sight of the storyteller's mouth resembles the core of an uprooted cedar trunk, his voicelessness recalling the trees' loss of language in the epic's opening scene. Like a tree, the storyteller is symbolically converted into a canoe, yet the border between land and sea is once more rendered porous. The rotting vessel of his torso will not remain seaworthy; instead, it will become a landlocked ruin decomposing in the soil underneath an almond tree. The image of the boys playing war or pretending to race in the "rotted hulk" projects this vision of the grounded canoe across the ocean into future generations that will survive the death of the African griot in the Caribbean, as evidenced by the presence of deciduous almond trees.[55] Like the seaborne captives who will arrive via the Middle Passage, Achille's vision traverses the Atlantic to become earthbound. The suggestion is that as the defunct canoe facilitates the play of generations of boys becoming men, it will also become part of the future forest, contributing its nutrients to the roots of the almond tree. Just as the canoe transported Achille supernaturally to his ancestors in an African village, it hyphenates his vision of the past back into the present. In figuratively becoming a canoe, the African griot can merge his body (and his stories) with the forest that will serve as a resource to the descendants of his enslaved kin. The hyphen-esque length of his body is overtaken by the hyphae of the decomposing fungi in the soil, and he becomes a vessel in more ways than one.

Notes

1 Derek Walcott and Peter Doig, *Morning, Paramin* (New York: Farrar, Straus and Giroux, 2016), 17.
2 Derek Walcott, *Omeros* (New York: Farrar, Straus and Giroux, 1990), 11.
3 Walcott and Doig, Peter, *Morning, Paramin*, 17.
4 Maria Cristina Fumagalli, *Derek Walcott's Painters: A Life with Pictures* (Edinburgh: Edinburgh University Press, 2023), 429.
5 George B. Handley, "A Postcolonial Sense of Place and the Work of Derek Walcott," *Interdisciplinary Studies in Literature and Environment* 7, no. 2 (2000): 17, https://doi.org/10.1093/isle/7.2.1.
6 Handley, "A Postcolonial Sense of Place and the Work of Derek Walcott," 17; and Elaine Savory, "Toward a Caribbean Ecopoetics: Derek Walcott's Language of Plants," in *Postcolonial Ecologies: Literatures of the Environment*, ed. Elizabeth M. DeLoughrey and George B. Handley (Oxford: Oxford University Press, 2011), 93, https://doi.org/10.1093/acprof:osobl/9780195394429.003.0004.
7 Elizabeth M. DeLoughrey, Renée K. Gosson, and George B. Handley, eds., *Caribbean Literature and the Environment: Between Nature and Culture* (Charlottesville: University of Virginia Press, 2005), 3.
8 Derek Walcott, "Isla Incognita," in *Caribbean Literature and the Environment: Between Nature and Culture*, ed. Elizabeth DeLoughrey, Renée K. Gosson, and George B. Handley (Charlottesville: University of Virginia Press, 2005), 57.
9 Walcott, *Omeros*, 56.
10 "Canoe" is believed to come from the Taíno word "kanawa." "Pirogue" and its variants are believed to come from the Carib language. Isaac Shearn, "Canoe Societies in the Caribbean: Ethnography, Archaeology, and Ecology of Precolonial Canoe Manufacturing and Voyaging," *Journal of Anthropological Archaeology* 57 (2020): 3, https://doi.org/10.1016/j.jaa.2019.101140; Julian Granberry and Gary S. Vescelius, *Languages of the Pre-Columbian Antilles* (Tuscaloosa: University of Alabama Press, 2004), 107; and Tessa Murphy, *The Creole Archipelago: Race and Borders in the Colonial Caribbean* (Philadelphia: University of Pennsylvania Press, 2021), 2, https://doi.org/10.9783/9780812299977.
11 Pedro Acevedo-Rodríguez, *Catalogue of Seed Plants of the West Indies*, ed. Mark T. Strong (Washington, D.C.: Smithsonian Institution Scholarly Press, 2012), 189; and Roger Graveson, "Plant Taxonomy of Saint Lucia: Botanical Descriptions of Important Species, Species Checklist, and Herbarium Development," *Technical Report No. 4 to the National Forest Demarcation and Bio-Physical Resource Inventory Project* (Helsinki: FCG International, 2009), 43.
12 Graveson, "Plant Taxonomy of Saint Lucia," 9; and "Saint Lucia Country Report to the FAO International Technical Conference on Plant Genetic Resources" (Castries: Ministry of Agriculture, Lands, Fisheries and Forests, 1995), 29, https://www.fao.org/fileadmin/templates/agphome/documents/PGR/SoW1/americas/SAINTLUC.pdf.
13 Shearn, "Canoe Societies in the Caribbean," 2.
14 Douglas MacRae Taylor, *The Caribs of Dominica* (Washington, D.C.: U.S. Government Printing Office, 1938), 141; and Shearn, "Canoe Societies in the Caribbean," 6.
15 Shearn, "Canoe Societies in the Caribbean," 2; and John Emmanuel, "The Traditional Gommier Canoe Race Is Here Again," Ministry of Public Service, Home Affairs, Labour and Gender Affairs, 2014, https://publicservice.govt.lc/news/the-traditional-gommier-canoe-race-is-here-again.
16 Édouard Glissant, *Poetics of Relation*, trans. Betsy Wing (Ann Arbor: University of Michigan Press, 1997), 6.
17 Glissant, *Poetics of Relation*, 6.
18 Pardis Mahdavi, *Hyphen* (New York: Bloomsbury Academic, 2021), 13.
19 Shearn, "Canoe Societies in the Caribbean," 10–12.
20 Benoît Bérard, Jean-Yves Billard, Thierry L'Étang, et al. "Technologie du fait maritime chez les Kalinago des Petites Antilles aux XVIe et XVIIe siècles," *Journal de la Société des Américanistes* 102, no. 1 (2016): 129–58, https://doi.org/10.4000/jsa.14688.
21 Walcott, *Omeros*, 3.
22 Natalie Melas, "Forgettable Vacations and Metaphor in Ruins: Walcott's 'Omeros,'" *Callaloo* 28, no. 1 (2005): 153, https://doi.org/10.1353/cal.2005.0028.
23 Anthony Carrigan, "Postcolonial Tourism, Island Specificity, and Literary Representation: Observations on Derek Walcott's *Omeros*," *Space and Culture* 13, no. 2 (2010): 155, https://doi.org/10.1177/1206331209358220.
24 Elizabeth M. DeLoughrey and Maria del Carmen Quintero Aguilo have argued for a tidalectic reading of *Omeros*. Kamau Brathwaite, *Conversations with Nathaniel Mackey* (Staten Island, N.Y.: We Press, 1999), 44; Elizabeth M. DeLoughrey, *Routes and Roots: Navigating Caribbean and Pacific Island Literatures* (Honolulu: University of Hawai'i Press, 2007), 47–48; and Maria del Carmen Quintero Aguilo, "The Sands of Un-Certainty: Tidalectically Synthesizing Nature and Culture in Derek Walcott's *Omeros*," *CEA Critic* 81, no. 1 (2019): 1–10, https://doi.org/10.1353/cea.2019.0001.
25 George B. Handley, "'The Argument of the Outboard Motor': An Interview with Derek Walcott," in *Caribbean Literature and the Environment: Between Nature and Culture*, ed. Elizabeth M. DeLoughrey, Renée K. Gosson, and George B. Handley (Charlottesville: University of Virginia Press, 2005), 129.
26 Handley, "'The Argument of the Outboard Motor,'" 129.
27 Walcott, *Omeros*, 3.
28 Mac Fenwick, "'All Strangers Here': 'Native' as Invasive in the Poetry of Derek Walcott," *Journal of West Indian Literature* 20, no. 1 (2011): 24.
29 The *laurier-cannelle* was previously identified incorrectly as *Cinnamomum elongatum*; however, recent research has determined that it is *Aniba ramageana*. R. de Kok, "Aniba Ramageana," in *The IUCN Red List of Threatened Species*, 2021, https://www.iucnredlist.org/species/143818770/143953136; and Roger Graveson, "Aniba Ramageana," in *Plants of Saint Lucia*, 2019, https://www.saintlucianplants.com/floweringplants/lauraceae/anibrama/anibrama.html.

30 Graveson, "Aniba Ramageana."
31 Shearn, "Canoe Societies in the Caribbean"; and Bérard et al., "Technologie du fait maritime chez les Kalinago des Petites Antilles aux VXIe et VXIIe siècles."
32 Fenwick, "'All Strangers Here,'" 24; and Walcott, *Omeros*, 3.
33 For more on plant communication, see Monica Gagliano, *Thus Spoke the Plant: A Remarkable Journey of Groundbreaking Scientific Discoveries and Personal Encounters with Plants* (Berkeley: North Atlantic Books, 2018).
34 Walcott, *Omeros*, 3.
35 Handley, "'The Argument of the Outboard Motor,'" 135.
36 Sneharika Roy, "Postcolonial Engagements with the Epics: Multiple and Movable Tectonic Plates," *Commonwealth* 34, no. 1 (2011): 30.
37 As Achille will observe, "The young took no interest in canoes / That was longtime shit." He will go on to make the point that the canoe also came from Africa; in this sense, it syncretizes African and Amerindian histories. Walcott, *Omeros*, 112.
38 Roger Graveson, "Cedrela Odorata," in *Plants of Saint Lucia*, 2019, https://www.saintlucianplants.com/flowering plants/meliaceae/cedrodor/cedrodor.html.
39 Walcott, *Omeros*, 5.
40 Walcott, *Omeros*, 6.
41 In *The Prodigal*, Walcott references the "sudden Babel of the manchineel." Derek Walcott, *The Prodigal* (New York: Farrar, Straus and Giroux, 2004), 62; and Acevedo-Rodríguez, *Catalogue of Seed Plants of the West Indies*, 435.
42 Savory, "Toward a Caribbean Ecopoetics," 92.
43 Walcott, *Omeros*, 5.
44 Handley, "'The Argument of the Outboard Motor,'" 135.
45 Walcott, *Omeros*, 5.
46 Walcott, *Omeros*, 5.
47 Walcott, *Omeros*, 4–5.
48 Walcott, *Omeros*, 6.
49 This term can refer to either *Schefflera morototoni* or *Cecropia schreberiana* in Saint Lucia. Acevedo-Rodríguez, *Catalogue of Seed Plants of the West Indies*, 69, 919.
50 Walcott, *Omeros*, 7.
51 Walcott, "Isla Incognita," 56.
52 Walcott, *Omeros*, 8.
53 Walcott, "Isla Incognita," 57.
54 Walcott, *Omeros*, 146–47.
55 While *Prunus africana*, sometimes known as the African almond tree, is found in tropical Africa, it is evergreen. Gerard Bodeker, Charlotte van 't Klooster, and Emma Weisbord, "*Prunus africana* (Hook.f.) Kalkman: The Overexploitation of a Medicinal Plant Species and Its Legal Context," *The Journal of Alternative and Complementary Medicine* 20, no. 11 (2014): 810–22.

Sacred

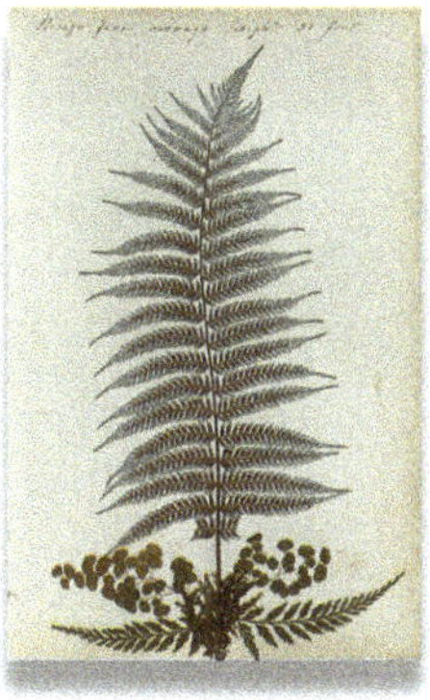

13

PLANTING BUDDHISM IN MEDIEVAL CHINA

NATASHA HELLER

In the context of Protestant Christianity, the phrase "church planting" refers to the establishment of new congregations where none had previously existed. It is a key part of missionizing endeavors and propagating the Christian faith. Were we to imagine a similar term to describe the spread of Buddhism, we might say that Buddhism was *enforested*. Trees abound in the narrative of the Buddha's life: he was born under a sāla tree (*Shorea robusta*), awakened under a bodhi tree (*Ficus religiosa*), and died between sāla trees (Figure 13.1). The centrality of trees continued in China, where Buddhists often referred to their monasteries as *conglin* 叢林, "dense forests."[1] Contrary to modern assumptions that groves or forests are larger than gardens, such monastic forests might be associated with gardens: we see this in the title of an important seventh-century Buddhist encyclopedia, *The Forest of Pearls in the Garden of Dharma* (*Fayuan zhulin* 法苑珠林). Indeed, the inclusion of trees within a garden that also holds buildings for worship and monastic residences becomes a dominant form of Buddhist placemaking from a very early point in China. Although scholarly attention has been almost exclusively directed toward the human-built portions of these monastic complexes, the "dense forests," along with other plants, also shaped the practice and perception of Buddhism. How Buddhism was enforested—whether with pomegranates or pines—mattered, establishing historical connections and inflecting ecosensoria.

Read with attention to trees and other plants, *A Record of Buddhist Gardens in Luoyang* (*Luoyang qielan ji* 洛陽伽藍記) demonstrates the diverse ways flora shaped medieval monastic life. *A Record of Buddhist Gardens in Luoyang* was written by Yang Xuanzhi 楊衒之, whose birth and death dates are unknown.[2] The text became part of the Buddhist canon and exists in two Ming dynasty editions.[3] The author is mentioned in the prosopographic collection *Further Biographies of Eminent Monks* (*Xu gaoseng zhuan* 續高僧傳), which is

FIGURE 13.1

Mounted bodhi tree (*Ficus religiosa*) leaves, 1640s. National Museum of Asian Art, Smithsonian Institution, Freer Collection, Gift of Charles Lang Freer, F1911.504a-t.

at the end of the biography of the famed translator Bodhiruci 菩提流支 (d. 527), who lived in Luoyang. There, Yang Xuanzhi is identified as the governor of the Qicheng 期城 Commandery. A brief excerpt from his introduction to *A Record of Buddhist Gardens in Luoyang* is appended to his biography.[4] Scholars have drawn on *A Record of Buddhist Gardens in Luoyang* for anecdotes about historical figures, descriptions of Buddhist architecture, layouts of monasteries, and literary history, as the text includes poetry and other snippets of literature. The text also depicts monastic life as deeply entwined with trees and other plants, which provided food, shade, and spaces conducive to contemplative practice.

Gardens as Religious Spaces

Monasteries and temples were often gardens—in China as in India. Gregory Schopen made this point nearly twenty years ago, arguing that early Buddhist monks "attempted to assimilate their establishments to the garden, or actually saw them as belonging to that cultural category."[5] In India, at the time that the earliest Buddhist institutions were being established, gardens were to be found on the outskirts of cities; they were embellished spaces that disguised the less attractive parts of nature. Indian gardens in the early centuries of the Common Era were characterized by walls, both to define the garden from the outside and to divide it into smaller spaces. Waterworks provided irrigation but also pleasing views through constructed ponds, meandering channels, and fountains. Bowers of trees offered shade and seclusion, and human-made hills allowed for taking in the views of the manipulated landscape.[6] The ideal garden was a springtime one, offering seductions both aesthetic and erotic. These qualities carried over to Buddhist monasteries,[7] and the garden dimension of these religious institutions remained important for those who visited them. Schopen puts it this way: "The *vihāra*—not the Buddha and the monks—is what they want to see, and these were in many instances, and from almost any point of view, quite an amazing sight."[8]

That Buddhist monks (and nuns) dwelt in gardens is obscured by how we typically refer to such places in English, which emphasize their human-built features. A "monastery" is defined as a building, with its etymological roots in aloneness. Indeed, "monastery" is how *qielan* 伽藍 in the title of the text under consideration in this chapter is usually translated. But *qielan* transliterates a Sanskrit term, *saṅghārāma*, meaning a garden or place of enjoyment.[9] Thus, a more accurate translation of *Luoyang qielan ji* would be *A Record of Buddhist* <u>*Gardens*</u> *in Luoyang*.

A garden in medieval China was necessarily quite different from one in India at the time that the historical Buddha is said to have lived, both in terms of flora and in the aesthetics that shaped garden-making. There was continuity in so far as gardens were the form or container for Buddhist institutions, but the meaning of that form was determined by its specific features and by the plants within it. A garden could take a number of sizes and forms, from small plots that supply a family's vegetables to vast imperial gardens and hunting parks. In addition to differences of scale, gardens varied by the social class of their owners. The Buddhist gardens described in *A Record of Buddhist Gardens in Luoyang* were most like those of the elites, and indeed were very often converted from secular gardens; lay Buddhists often turned over their estates to enable the founding of a temple or monastery. These elite gardens were perhaps somewhat smaller in size than those belonging to the imperial family but might equal them in inventiveness and luxury.

Gardens were subject to social and cultural forces, and the medieval period represented an epochal change from early designs. The features that characterized medieval gardens would come to define gardens in later periods as well; to put it another way, the medieval gardens of Luoyang and other wealthy urban centers were tremendously important to the history of gardens in China. There were two key changes. First, private gardens (as opposed to imperial ones) assumed a new significance. Second, gardens began to imitate the natural landscape. As Wendy Swartz has noted, views of nature shifted from "inhospitable and dangerous" in the early period to a "safe haven from the world of affairs," which is the view the poet Xie Lingyun 謝靈運 (385–433) took when he wrote his "Rhapsody on Dwelling in the Mountains" (Shanju fu 山居賦).[10] Landscape, as Xie and other scholar-artists of this period recognized, could do more than provide a retreat from official life; it could also furnish spiritual experiences. The painter Zong Bing 宗炳 (375–443) wrote that "landscapes give form to the beauty of the Way, and humane people delight in them."[11] Looking at landscapes could be a form of ethical cultivation and cosmological attunement. Zong Bing recognized that it was not always possible to visit real landscapes, and so painted ones could stand in, allowing the viewer access to similar effects.

Gardens, likewise, came to recreate nature. The imperial gardens in the Qin and Han dynasties had grand artificial lakes; by the medieval period, the ideal had become naturalistic waterscapes. Gardens situated on flat terrain were transformed through human effort to offer varied views. The goal was to construct a space that felt like being in nature, and gardens might indeed incorporate parts of mountains.[12] The suburbs of Luoyang had a cluster of such private gardens, among them that of Zhang Lun 張倫, which was one of the most spectacular. He built an artificial hill, creating a series of views that mimicked a remote retreat. It featured a path that seemed obstructed but was not, along with trees so dense that they blocked light. While some criticized Zhang for his lavish lifestyle, which was manifest in an expansive garden that tried to recreate nature, others saw it as a praiseworthy endeavor.[13] In short, elite gardens in medieval Luoyang imitated nature to provide their owners with aesthetic and spiritual enjoyment.

Monastic gardens had special requirements because they had both practical and aesthetic roles in Buddhist life. Chu Zhaowen 儲兆文, in his modern *History of Chinese Gardens* (*Zhongguo yuanlin shi* 中國園林史), identifies three environmental factors in the siting of medieval monastic gardens. First, they needed to be near water, as that was a daily necessity. Second, gardens should be located near forests, which provided both pleasing landscapes and material for building and firewood. Third, they should be in a favorable microclimate (*xiaoqihou* 小氣候), cool but sheltered from the wind and with good sun exposure.[14] Taken together, these factors suggest the practical functions that gardens served. To his list, we can add spiritual cultivation and sociality. Plants play clear, and clearly different, roles in all three dimensions of Buddhist gardens.

Before turning to the types of plants that feature in the gardens described in *A Record of Buddhist Gardens in Luoyang*, it is important to get a sense of the city in which they are located. Luoyang is located on the banks of the Luo River, in what is now the province of Henan 河南. During the century under consideration here, this part of China was in a cold climatic stage.[15] It was a capital city for several dynasties and the location for many early projects to translate Buddhist scriptures into Chinese. The popularity of Buddhism in this region is attested not only by the temples within the general confines of the city but also in the cave complexes at Longmen 龍門, several miles south of Luoyang. There, monumental statues, memorial niches, and extensive devotional activities attest to the social and economic impact of Buddhism.

Documenting Buddhist Sites

In his introduction, Yang gives a brief history of Buddhism in China, beginning with Emperor Ming of the Han. He writes that in the Yongjia 永嘉 era of the Jin 晉 dynasty (307–313 CE), there were only forty-two Buddhist temples in Luoyang, presumably before its sacking by the Xiongnu 匈奴. After the Wei 魏 dynasty established the city as the capital in 495 CE, Buddhism flourished in all dimensions, from temple-building to statue-making, from personal conversions to sūtra-copying. This remarkable period lasted only a few decades, and the city was destroyed by the military commander Gao Huan 高歡 (496–547) in 534 CE. Yang says that just prior to this there were over a thousand temples. Yang is prompted to set down his record by a visit in 547 CE: "The outer and inner city walls had fallen in, palaces had collapsed, Buddhist and Daoist temples were in ashes, shrines and stupas were no more than mounds of ruins. The walls were covered with weeds, and the lanes filled with brambles. Wild beasts made burrows in abandoned stairways, and mountain birds nested in the trees of the courtyards. Wandering children and shepherds loitered in the thoroughfares; farmers and ploughmen planted millet on the grounds of the two watchtowers."[16] It is this sad state that prompts him to his efforts: "Now the [temples] are all empty and the sound of their bells seldom heard. I fear that their histories would not be transmitted to later generations, so I have set down this record. The temples were so many, however, that I could not describe them all; for the present I have confined myself to the great *saṅghārāmas*, including medium or small temples only when there were auspicious or strange events, or worldly stories, associated with them."[17] Framed by the abandoned sites of the city, Yang's work takes on an elegiac aim that implies a positive attitude toward Buddhism. His text proceeds from those temples within the city proper to those in its suburbs, organized by direction. There are differences of opinion as to whether the text is all of a piece or whether Yang intended to have main entries supplemented by notes or commentary.[18] Yang's visit to Luoyang was over a dozen years past Gao Huan's invasion. He, therefore, works from memory, not from what he observes as he writes. The resulting record is, as Manling Luo puts it, "what he believed to have happened and how he understood the significance of those practices."[19]

Yang uses various terms for religious institutions in his preface. He mentions *si* 寺 and *guan* 觀, which usually refer to Buddhist and Daoist temples or monasteries, as well as *miao* 廟, perhaps smaller shrines, and *ta* 塔, stupas—all different types of buildings that have fallen into ruin. These, however, are not the focus of *A Record of Buddhist Gardens in Luoyang*, and when Yang identifies what he includes, he uses the word *qielan* or *saṅghārāma*, modifying it by "great" (*da* 大) and signaling that the largest and most important Buddhist establishments in Wei dynasty Luoyang were gardens. Note, too, how the nonhuman appears in the passage quoted above: animals nest in places that would have formerly been off-limits or inhospitable to them, while weeds and brambles quickly take over, remaking the human-built world. Buildings, when abandoned, are often robbed of their most valuable parts and, if not maintained, swiftly fall into ruin—but the same is not necessarily true of trees and plants. The birds nest in the trees in the courtyards, after all, and those trees are still there for Yang to observe.

Manling Luo is one of the few scholars to treat *A Record of Buddhist Gardens in Luoyang* as literature in its own right, rather than as a source of historical facts and anecdotes. Luo argues that Yang Xuanzhi "foregrounds three major modes of place-making," identifying these as architectural, literary, and hermeneutic. The first two are straightforward: people make places by erecting or modifying buildings, as well as by writing about them. The hermeneutic mode of placemaking brings current spaces into dialogue with the past, and Yang does this by "identifying and interpreting the histories of specific structures and sites."[20] Plants are also part of placemaking, both through human intervention—as people bring trees, shrubs, and smaller plants into built spaces—and through their own activity. As we have seen in the description of a ruined Luoyang, plants grow without human intervention and may do so in unpredictable ways. Given that medieval gardens were in proximity and sometimes carved out of forests, there is also a dialogue in Buddhist gardens between plants within and outside their boundaries.

A Sacred Tree

Some of the trees in *A Record of Buddhist Gardens in Luoyang* are individuated. The mulberry tree (*sangshu* 桑樹) of Yuanhui Temple 願會寺 is one such example. This is likely a smaller temple, as it does not merit its own entry, and, like many of the other temples in Luoyang, it was converted from an estate donated by an eminent literatus. Part of its transformation was the building of a Buddha Hall (*fotang* 佛堂), in front of which was a mulberry tree (*sangshu*). This was not just any mulberry tree, however: "It grew five feet straight up, its branches spreading horizontally all around with twigs and leaves arrayed so that it looked like a parasol. Five feet higher than this, it was the same. Altogether there were five levels, and the leaves and fruits of each level were different."[21] This is not the end of the account of the tree, but let us pause for a moment to marvel and puzzle at it. Has this tree taken on such an intriguing form naturally, or through human intervention? The text does not mention anyone *doing* things to the tree. Leaves on the same tree can look different, based on where they are located height-wise on the tree and how much light they get. There are different kinds of mulberries with fruits of different colors; the fruits might have ripened at different times. If there had been human intervention, it might have taken the form of grafting. There are claims that the practice of grafting in China dates to before the beginning of the Common Era, but more reliably it is attested in *Essential Skills of the Common People* (*Qimin yaoshu* 齊民要術), an agricultural text roughly contemporaneous to *A Record of Buddhist Gardens in Luoyang*. It is possible to graft together multiple trees, which could explain the Yuanhui Temple mulberry's distinctive five layers. Depending on how literally we take the description of those layers as being like parasols or umbrellas, such shaping might be explained as topiary.

Imagining this mulberry as topiary or as grafted may be no more than a speculative exercise, but we should not overlook the fact that people associated with the temple almost certainly had to intervene in plant lives on occasion. The wonderment at the Yuanhui Temple mulberry does not depend on whether the tree is naturally this way or has been created through human intervention. Whatever its source, the aesthetic of the tree is distinct: unlike garden views constructed to look as if they were natural and blending in with the landscape beyond, this mulberry stands out for its unusual, perhaps even supernatural, shape. This shape might well have resembled a pagoda or stupa, or part of a stupa. For example, medieval pagodas often had five levels. Stupas were topped with spires, to which were added discs called *chatra* or parasols. Such visual resonances would have added to the mulberry's specialness.[22]

Yang includes the remarkable mulberry because it became a highly popular local attraction. Picking up where we left off in the passage, we read:

> Both monks and lay-people called it a "sacred mulberry." The crowds who came to see it were like those in the marketplace; donors were also very numerous. The emperor [Xiaowu of the Northern Wei] heard about it, and despised it, worried about what might happen with such crowds. Thus he ordered Yuan Ji 元紀, Palace Secretary and Attendant at the Yellow Gate, to cut it down. On that day clouds darkened the sky, and from the place where the axe landed, blood flowed out onto the ground. All those who saw this could not help but shed sad tears.[23]

The unusual features of the tree led people to refer to it as "sacred"—not only did it draw spectators but it also inspired donations. Given that he was later driven out of Luoyang, the young Emperor Xiaowu (b. 510, r. 532–535) was perhaps not wholly wrong to be concerned with crowds. Nor did he trust the task of chopping down the mulberry to a minor workman, sending instead someone of standing to do the job. The popularity and sacred nature of the tree is a matter of some importance, but its location at a Buddhist temple is not incidental. Emperors were often worried about the alternative claims to authority religion might make, and extravagant donations were an indicator of popular enthusiasm. Lay expenditures inspired by this tree drew the emperor's ire much as the popularity of other Buddhist objects did in other contexts.

The description of the dark skies on the day the tree is to be chopped down might be read as an omen, although Yang does not explicitly do so. As for the description of blood flowing out of the tree, there are some species with red sap, but mulberry is not one of them and should have whitish sap. Bleeding trees, however, are found elsewhere in medieval literature. Among the several anecdotes related to trees in the fourth-century *Record of Searching for the Supernatural* (*Soushen ji* 搜神記), there are two examples of bleeding trees. In one case, after blood pours from a tree, the tree falls and strange creatures run out of it. In the other, the retainer sent to chop down a troublesome catalpa—its canopy shading the wrong place—is terrified by the blood pouring out of the tree and runs to tell his boss, the local governor. Not intimidated by strange happenings, the governor finishes the job.[24]

Whatever the explanation, blood is an animating force, and in the *Record of Searching for the Supernatural*, it is one way of marking some trees as special. Blood plays a similar role here, attesting that the mulberry was not just a tree but something more. What made it *shen* also made it bleed and turned it into an object of mourning. It is worth noting how this particular tree differs from others we will encounter—which could be described as anonymous, undifferentiated trees. The mulberry, on the other hand, is singular, and its individuating features make it special.

Buddhist and Botanical History

Another exciting, exceptional tree (or trees) highlighted in Yang's *A Record of Buddhist Gardens in Luoyang* is found at Baima Temple 白馬寺. This was purportedly the earliest Buddhist temple to be established in China; it was named after the white horse that carried back scriptures from the mission sent by Emperor Ming 明帝 of the Han dynasty (r. 57–75 CE). The Silk Road associations of the Baima Temple's origin story are continued in its flora. In front of the stupa, there were grapes "different from other places" and pomegranates. Both of these plants had been introduced to China from Central or Western Asia, and both are associated with the general Zhang Qian 張騫 (d. 114 BCE), whose travels to western regions led to the introduction of these and other fruits. Zhang Qian undertook his mission at the order of Emperor Wu 武帝 of the Han (r. 141–87 BCE), who wished to establish relations with the Yuezhi 月氏. It was not a short or simple journey: Zhang was captured by the Xiongnu and spent ten years with them, but following this delay, he passed through Dayuan 大宛 (or Fergana, in what is now eastern Uzbekistan). It is there that he collected grape seeds, as well as alfalfa seeds. Both were planted successfully in China, leading later envoys to comment on how well they grew.[25] Of course, grapes were not cultivated primarily as fruit for raw consumption. In the contexts in which Zhang encountered them, they were used for making wine. As Robert Spengler writes, "grape wine became one of the most important commodities of the ancient world, allowing people to accumulate an agricultural surplus that could be stored for decades and to display their wealth."[26] Spengler notes that grapes were also dried, which was perhaps a mode of preservation more appropriate for a Buddhist site.

The grape is here paired with another Silk Road fruit: *tulin* 荼林, translated as "pomegranate." Language is one way that the historical movements of plants can be traced. The anthropologist and historical geographer Berthold Laufer (1874–1934) notes that *tulin* seems connected to the Sanskrit term for pomegranate, *dāḍima* or *dālima*; however, it is not explained in Chinese sources as being a Sanskrit term, and *tulin* and its variant *tulin* 塗林 do not appear in historical Chinese-Sanskrit dictionaries. Given that *tulin* was associated with Zhang Qian, Laufer speculates that it reflects a now-lost term for "pomegranate" in an Iranian language. The etymology of the more common term for pomegranate, *anshiliu* 安石榴 (or *shiliu* 石榴), is no less clear, but Laufer speculates that *liu* transcribes an Iranian word and that *anshi* refers to Parthia.[27]

Grapes and pomegranates would have likely entered China before Buddhism did, although perhaps not by much. That they were planted at Baima Temple reinforces the associations of that Buddhist institution with the developing Silk Road and regions to the west, through flora on the temple grounds evoking other geographic locations. (As an aside, we might see the grapes and pomegranates at this temple as a very small collection of the type associated with botanical gardens.) Their significance was not limited to their mere presence, because pomegranates and grapes grew exceptionally well within the bounds of the temple garden:

> Their branches and leaves were abundant, and their fruits were exceptionally large. The pomegranate fruits weighed seven pounds, and the grapes were bigger than jujubes.[28] Their flavor was also particularly fine, the best in all of Luoyang. When they ripened the emperor would come especially to pick them. Sometimes he would then give them to the palace concubines. When the concubines received them, they would in turn present them to their relatives. Because of their unique flavor, those who were given them did not dare eat them at once, and so they passed through several households. There was a saying in the capital, "The sweet pomegranates of Baima are each worth an ox!"[29]

The emperor mentioned in this passage is likely Emperor Ming, to whom the introduction of Buddhism to China is credited. That he was picking pomegranates at the site of Buddhism's founding in China is a reminder of how plants became bound up with the movement of peoples and ideas—and how remarkable and wondrous new plants and fruits would have seemed. To illustrate this wonder: Baima Temple had scripture cases that occasionally radiated light, causing both monks and laypeople to venerate them as if they were the image of the Buddha. This astounding feature merits one sentence, whereas the flavor, size, and social importance of the pomegranates are described in greater detail. The pomegranates, as they passed from household to household, were juicy envoys for the temple. The wonder that the emperor and his associates felt toward the grapes and pomegranates was a positive emotion. The sacred mulberry, on the other hand, shows how unusual or marvelous plants could also represent a challenge, growing beyond official control.

Pomegranates serve as a kind of proxy for Buddhism. The fruit, from points west, was able to grow with great success in a new place, akin to Buddhism's flourishing (as attested by Yang's work). As the fruit is passed from elite household to elite household, we can imagine that its origin was recounted to each new recipient, a physical parallel to the way Buddhist ideas and practices may have moved—their adoption prompted by social connections. Pomegranates also have another kind of symbolism: their multitudinous seeds were often likened to sons. Fertility meant posterity.

More broadly, the metaphor of fruit is important in Buddhism: the results of practice are known as "fruits" (*guo* 果). Gardens create spaces for both kinds of fruit, real and metaphorical. Indeed, the successful growth of real fruits may have reflected attainments of the fruits of the Buddhist path, insofar as monastic grounds may have accrued a kind of supernatural potency from the practices of monks and nuns. Moving in the other direction, the consumption of real fruit was part of the nutritional support offered to religious practitioners in residence.

The sacred mulberry and the prolific pomegranate are individuated trees (even if there were multiple pomegranate trees), special in some way. They are also caught up in narratives about important individuals—emperors, in both cases—and are examples of what Manling Luo has called "hermeneutic place-making," here taking place through trees. After all, the mulberry is not there for Yang to observe, having been cut down many years before. Scholars of landscape studies might call these witness trees: these are usually large, old trees that are around for key moments of history and that often persist in cultural memory even after they are gone.

The pomegranate and mulberry trees answer the question of what makes a garden Buddhist. In one case, it is a Silk Road tree that found a new and fertile home in China; in the other, it is a tree whose unusual and marvelous growth attests to the potency of Buddhist places in a different way. Turning from Buddhist transformations of plants, I want to reverse direction, looking at how plants might have shaped Buddhism. In other words, how do garden sites alter what Buddhism is in medieval China?

Shade and Fruit

Most of the trees in *A Record of Buddhist Gardens in Luoyang* are not individuated. Yet more quotidian trees—let alone plants—growing in multiples, one not readily distinguished from the rest, are no less important for understanding how flora has shaped places, including the religious sites of Luoyang. Although individual trees with special histories may seem more compelling, undifferentiated trees are arguably more *important*. Such trees are in the majority in monastic gardens, and they do more to shape the spaces of everyday practice therein. They are cocreators of Buddhist life.

Trees are first mentioned in the first entry, a description of the Yongning Temple 永寧寺, established by Empress Dowager Ling 靈. While there are plants within the boundary walls of this monastery, arboreal encounters also occur beyond its borders: "Outside all four gates are planted trees such as green scholar-trees, extending along clear water. Many travelers to the capital took shelter beneath them. On the street, the floating dust was cut off; this was not due to the moisture of the rising clouds. And the coolness brought by a refreshing breeze, how could this be produced by a silk-tree fan?"[30] What is implied by the final two sentences is that the environment just outside the monastery gates provides respite from the burdens of travel. (A minor point—the round fan, either made from leaves or resembling them, indicates how plant products were integrated into all areas of life.) Scholar-trees (*Styphnolobium japonicum*) are often fifty feet tall, with a wide canopy and summer blossoms. Someone coming to Luoyang would first encounter the trees, perhaps take rest beneath them, and, through such an experience, form associations between shade and Buddhist monasteries.

Similar experiences could be found in proximity to other Buddhist establishments. The Jingle Temple 景樂寺 bordered an urban ward in which a well provided for the use of all (a "well of righteousness," *yijing* 義井) was located. North of this ward were "several mulberry trees with luxuriant branches; beneath them was a well of sweet water. There was a stone water trough and metal cups for the use of travelers, who drank the water and took shelter in the shade of the trees; many took a rest there."[31] Ordinary mulberry trees like these generally grow between thirty and fifty feet tall and have a dense canopy. In addition to bearing fruit, they are perhaps best known in the Chinese context for their leaves, which are the preferred food of silkworms. A row of mulberry trees would indeed provide ample shade, but it would also carry with it connections of a society that provided for all: the philosophical text *Mencius* claims that planting mulberry trees will allow the elderly to wear silk, a sign of a well-ordered and compassionate state.[32]

Immediately following this comment on the shady well, Yang launches into a description of the temple buildings at Jingle. Although the well and the mulberry trees are outside—and, indeed, a little bit away from the temple proper—Yang's account places them together. This signals the qualities connected with Buddhist temples: they are literal and metaphorical places for shelter and rest.

The grounds of Buddhist sites had to be at least partially functional, although such practical planting might be out of view or deemed of less interest by men like Yang Xuanzhi. Occasionally, there are glimpses of how plants helped provide for the monastery. In the entry for Longhua Temple 龍華寺, Yang mentions two nearby temples, the Zhuisheng Temple 追聖寺 and the Baode Temple 報德寺. About these three temples, he observes: "The temples of the capital all planted various fruit trees, but the gardens of these three monasteries were the most productive."[33] The garden of the Jinglin ("Forest View") Temple was filled with "rare fruit" (*qiguo* 奇果), although we do not learn exactly which ones.[34] In a ward near the Baode Temple, there are more orchards associated with temples: "In this ward there were the three temples named Wenjue 文覺, Sanbao 三寶, and Ningyuan 寧遠. All around them were gardens that produced precious fruits, such as the Hanxiao 含消 pears from Dagu 大谷, which weighed ten pounds and turned into juice if they fell from the tree onto the ground. People of the time said: 'The pears of Baode 報德 are like the crab-apples of Chengguang 承光.' Chengguang Temple also had many fruit trees, and their crab-apples were particularly flavorful—the best in the capital."[35] Buddhist gardens produced fruit so good and flavorful that it was the talk of the town, with names—if not cultivars—that reflected the temple where they were grown.

Plants and Practice

Drawing closer to the main buildings of these temples, the impressiveness of the living quarters for monks and nuns is echoed in the abundance of plant life that surrounds them. At last returning to Yongning Temple, we read: "The dormitories for monks were imposing buildings, with over a thousand rooms. They had incised beams and whitewashed walls; the doors had blue patterns, and the windows had carved designs. It is difficult to put into words! Juniper and cypress, pine and cedar—their branches brushed the eaves. Thickets of bamboo and fragrant grasses surrounded the stairways. Chang Jing 常景 (d.u.) wrote a stele that reads, 'The main hall on Mount Sumeru and the cloisters in Tuṣita Heaven cannot surpass this!'"[36] In this brief description, the marvelous architectural details find parallel in the profusion of trees. These trees differ from those that provided shade for weary travelers: four types of evergreens surround the halls, perhaps matched in height by the bamboo.

The Yaoguang Temple 瑤光寺 likewise, had dormitories that were surrounded by greenery: "The lecture halls and nuns' rooms numbered over five hundred, with finely carved doors and windows one after another; precious trees and fragrant grasses were beyond description. Beef-tendon trees and dog-bone trees, chicken-head grass and duckfoot grass were all there." The first two types of trees have been identified as evergreens and hollies.[37] Yongning Temple and Yaoguang Temple were not the only Buddhist establishments where we read of evergreens. The Jingming Monastery 景明寺, founded south of the city by Emperor Xuanwu 宣武帝 (r. 499–515 CE) of the Northern Wei, was "shaded by green groves" in a beautiful landscape. "Beyond the eaves," we read, "were mountain forests and ponds. Pines and bamboo, orchids and angelica all lined the stairways; their fragrance was carried on the wind and condensed into dew."[38] The pairing of evergreens and bamboo recurs in the description of the Yongming Temple 永明寺, also in the western suburbs, and established by Emperor Xuanwu. Imposingly large, with thousands of rooms, "in the courtyard were arrayed tall bamboo, and the eaves were brushed by lofty pines."[39] The juxtaposition of plant life and architectural elements—with trees that touch the eaves and surround the stairways—points to proximity: the pines are not *outside* the monastery, but right there. Why would it be significant to have these particular trees near where the monks lived?

A hint comes in the description of Ningxuan Temple 凝玄寺: "Bamboo and cypress formed groves—it was truly a place of pure practice and stopping the mind," that is, maintaining celibacy and meditating.[40] Here, the pairing of evergreen and bamboo is linked to specific kinds of Buddhist behavior. Bamboo is mentioned alongside pine or cypress frequently enough for us to think that this is a deliberate choice. Of course, we cannot know if such choices are, indeed, part of the gardening practices of the Buddhists of medieval Luoyang or if Yang Xuanzhi chose them to make a point—but given how common all three plants are, there is no reason to suppose that Buddhist gardens did *not* have pines, cypresses, and bamboo planted near each other.

These three trees all stay green throughout the year, even during the cold of winter.[41] Pines and bamboo, along with the other evergreen trees mentioned, thus have associations with longevity, perseverance, and integrity. Pines can survive being buffeted by the wind, and bamboo bends without breaking; these qualities became metaphors for how a gentleman should endure the challenges of life, especially its political dimensions. By the Song dynasty, pine and bamboo had become associated with the plum tree, and they were together known as the "three friends of winter." Their collective symbolic associations were already developing during this earlier period.[42]

The qualities that the pine and bamboo represent for the scholar-official could be transferred, in part, to monks or nuns. Their lives are defined by constancy, maintaining the precepts, and engaging in meditation, without regard for external conditions. If the plum were at this point already associated with pine and bamboo, it is perhaps not surprising that it would be left out of plant symbolism for Buddhist clergy: a plum blossom is also connected to springtime and beauty, less readily associated with monastic ideals. Pine and bamboo, however, embody literati virtues that align with monastic ones. Whereas Indian gardens privileged the aesthetics and erotics of spring, Chinese Buddhist gardens created new webs of meaning through plant life.

In the description of Ningxuan Temple, cypress and bamboo are paired with what might be considered paradigmatic Buddhist practices. We have also seen passages in which the monks' rooms are surrounded by flora. This is true of other places of cultivation as well: Jinglin Temple describes a meditation chamber (*chanfang* 禪房) in the midst of a garden: "Within was set up a *Jetavana-vihāra*, and although it was quite small, its clever construction was without compare. Moreover, the meditation pavilions were silent, and the contemplation rooms were remote, with fine trees close up against the windows, and fragrant pollia surrounding the staircases. Although in the midst of the city, it seemed like being in a mountain valley. Monks who practiced quiescence had rope seats within; eating the wind and wearing the Way, they sat crosslegged and counted their breaths."[43] *Pollia japonica* is a flowering deciduous plant, and, along with the unnamed trees, it helps to create an atmosphere conducive to meditation.[44] The plant life of Jinglin Temple formed a visual and aural screen between the meditation rooms and the outside bustle of markets and court activities. As a consequence, monks could undertake austerities and contemplative exercises more in line with eremitic sites. At the Qin Taishang Jun Temple 秦太上君寺—named after the mother of the empress dowager—plant life abuts practice spaces: "The recitation rooms and meditation halls were on all sides, one on top of another. Flowering groves and fragrant grasses surrounded the staircases. There were often famous and eminent monks who lectured on Buddhist texts, and śramaṇas who had received the precepts there numbered in the thousands."[45] We are not given information here about the effects of the trees and grasses, yet they appear alongside descriptions

FIGURE 13.2
Ink painting on a bodhi tree (*Ficus religiosa*) leaf, 1640. National Museum of Asian Art, Smithsonian Institution, Freer Collection, Gift of Charles Lang Freer, F1911.504a-t.

that emphasize the success of the temple in multiple domains of Buddhist practice (chanting, meditation, lectures) (Figure 13.2).

For plants to best create an atmosphere conducive to religious practice, they needed to be cultivated with seasonality in mind. Dajue Temple 大覺寺, located in the western suburbs, was once the residence of a prince, which might account for its spectacular views of both mountains and rivers. The temple grounds were also planted to provide different views across the seasons: "When spring arrived the wind shook the trees, and the orchids opened purple petals; when the autumn frost descended on the grasses, the chrysanthemums sent forth yellow flowers. Famous monks and great worthies here dispatched their attachments through quiescence."[46] Again, this juxtaposition is instructive: plants create environments that are conducive to the most essential types of Buddhist practice.

Religious practice is facilitated by patronage—here, well-planted gardens also had a role. The sacred mulberry tree at Yuanhui Temple showed how monastic gardens could become popular attractions. This was true for other sites too. In the western suburbs of the city, Baoguang Temple 寶光寺 sat on the site of an earlier temple, whose fruiting trees and vegetable garden were still remembered by some. In Yang's time, Baoguang Temple also had a garden: "In the garden there was a lake called 'All-Encompassing Pool.' Rushes and reeds covered its banks; caltrops and lotuses blanketed the water. Deep green pines and emerald bamboo spread along its sides." This garden created such a pleasant space that officials of the capital would gather there, often in great numbers, to enjoy themselves in such pursuits as wine drinking and poetry writing.[47] The sightseeing potential of Buddhist gardens is a point of continuity with their South Asian antecedents. In another temple, converted from an estate, visits were especially associated with the Buddha's birthday:

> On the eighth day of the fourth month, men and women of the capital came in great numbers to the Hejian Temple 河間寺; after seeing the beautiful cloisters all sighed, saying that the dwellings of the immortals on Penglai could not surpass them. When they entered the back garden, they saw the winding irrigation channels, and stone stairs soaring upwards; red lotuses sprung from the ponds, green duckweed floated on the water. The pavilions were connected by suspended beams, and tall trees reached through the clouds. Everyone clamored over the garden, saying that even the Rabbit Garden of the Prince of Liang could not match it.[48]

While the account ends here, we might speculate that providing sensational—in both senses of the word—gardens for the wealthy to admire would have helped generate donations and other kinds of support.

Conclusion

The religiously advantageous environment of the Buddhist garden was a multisensory one, hinted at in the color of the lotus and duckweed as well as in the description of the Dajue Temple discussed above. The visual dimension is clearest in Yang Xuanzhi's description, with the different colors of the orchids and chrysanthemums lending special qualities to each season. Buddhist clergy and visitors also would have heard the trees. When we read that the wind shook them, we can imagine the sound of the leaves trembling against each other. Tactile experiences of plants would have been more active on the part of human residents. More than one passage mentions the plants surrounding the staircases, and residents might have brushed against their branches or reached out to touch a thick bamboo leaf or the bark of a tree trunk. Fragrance was also an important component of what plants contributed to Buddhist gardens; the monastery was *xiang* 香 in many ways, and, of course, *xiang* is also the word for incense. At the Jingming Temple, for example, the scent of pines, bamboo, orchids, and angelica permeated the site. Thinking of the use of aromatics in rituals, sweet-smelling plants contribute to an atmosphere of ritual purity. The sharp smell of pine, the softer scent from a juniper, or the aroma from a cedar's loose bark would have had similar effects. For flowering trees, scented blossoms turn into fragrant fruit, and Buddhist gardens often had fruit trees or were close to them. Buddhism had the taste of sweet fruits that grew well in monastic orchards.

A Record of Buddhist Gardens in Luoyang was not intended to provide us with an understanding of plants in Buddhism. Yang Xuanzhi's primary agenda is elsewhere. However, a careful reading shows that trees and other flora were indeed important parts of Buddhist life. Shifting our attention in texts is much like shifting our attention in life, from walking by plants to seeing them as central. When we change our focus in this way, we see how plants cocreated medieval Buddhism. How did they do so? They could be reminders of the foreign nature of the religion, living versions of statues from distant lands, and equally precious. They could provide sustenance and spectacle, refuge and rest. They created a sensory experience, and they facilitated contemplative practices. Trees and plants also demand care and cultivation, and individual specimens, such as the sacred mulberry, could come to mean more than their vegetal status might seem to imply. Plants were an essential part of placemaking throughout Luoyang, and they did so not entirely within human control, unlike architectural, artistic, or literary endeavors. Yang Xuanzhi recollects a recent past of Buddhist flourishing (a word that has its roots in "flower"), and as Buddhism came to be planted in medieval China, it was alongside flora of many types. These were no less part of Buddhist gardens—which is what Luoyang's great monasteries were—than grand statues and soaring architecture. The junipers and cypresses brushed the eaves, after all, and their presence shaped the practice beneath them.

Notes

1 This is a translation of the Sanskrit term *piṇḍavana*.

2 Not only that, but there also is some question about which character should be used for his surname; indeed, one text changes the character midstream.

3 Manling Luo, "The Politics of Place-Making in the *Records of Buddhist Monasteries in Luoyang*," *T'oung Pao* 105, nos. 1–2 (2019): 43n1. Based on quotations in other sources, the text circulated under different titles.

4 Takakusu Junjirō 高楠順次郎 and Watanabe Kaigyoku, eds., *Taishō shinshū daizōkyō* 大正新修大藏 (Tokyo: Taishō issaikyō kankōkai, 1924–1932).

5 Gregory Schopen, "The Buddhist 'Monastery' and the Indian Garden: Aesthetics, Assimilations, and the Siting of Monastic Establishments," *Journal of the American Oriental Society* 126, no. 4 (2006): 489.

6 Daud Ali, "Gardens in Early Indian Court Life," *Studies in History* 19, no. 2 (2003): 231–33, https://doi.org/10.1177/025764300301900204.

7 Schopen, "The Buddhist 'Monastery' and the Indian Garden," 490–94.

8 Schopen, "The Buddhist 'Monastery' and the Indian Garden," 496.

9 Schopen, "The Buddhist 'Monastery' and the Indian Garden," 487.

10 Wendy Swartz, "There's No Place Like Home: Xie Lingyun's Representation of His Estate in 'Rhapsody on Dwelling in the Mountains,'" *Early Medieval China* 21 (2015): 22.

11 Susan Bush, "Tsung Ping's Essay on Painting Landscape and the 'Landscape Buddhism' of Mt. Lu," in *Theories of the Arts in China*, ed. Susan Bush and Christian Murck (Princeton: Princeton University Press, 1983), 145.

12 Chu Zhaowen 储兆文, *Zhongguo yuanlin shi* 中国園林史 (Shanghai: Dongfang chuban zhongxin, 2008), 62, 66.

13 Yi-t'ung Wang, trans., *A Record of Buddhist Monasteries in Lo-yang* (Princeton: Princeton University Press, 1984), 91–93; and Luo, "The Politics of Place-Making in the *Records of Buddhist Monasteries in Luoyang*," 61.

14 Chu, *Zhongguo yuanlin shi*, 75.

15 Zhixin Hao, Jingyun Zheng, Yingzhuo Yu, et al., "Climatic Changes During the Past Two Millennia Along the Ancient Silk Road," *Progress in Physical Geography* 44, no. 5 (2020): 615.

16 Yang Xuanzhi 楊衒之, *Luoyang qielan ji jiaojian* 洛陽伽藍記校箋, annotated by Yang Yong 楊勇 (Beijing: Zhonghua shuju, 2006), 1–2. All translations are my own unless otherwise indicated. Cf. Wang, *A Record of Buddhist Monasteries in Lo-yang*, 6; and W. J. F. Jenner, *Memories of Loyang: Yang Hsüan-chih and the Lost Capital (493–534)* (Oxford: Clarendon Press, 1981), 142.

17 Yang, *Luoyang qielan ji jiaojian*, 2. Cf. Wang, *A Record of Buddhist Monasteries in Lo-yang*, 7; and Jenner, *Memories of Loyang*, 142.

18 As what are taken by some editors to be notes are not identified by textual markers, I have treated all the text as equivalent in status. See the discussion of this issue in Luo, "The Politics of Place-Making in the *Records of Buddhist Monasteries in Luoyang*," 47.

19 Luo, "The Politics of Place-Making in the *Records of Buddhist Monasteries in Luoyang*," 47.

20 Luo, "The Politics of Place-Making in the *Records of Buddhist Monasteries in Luoyang*," 48.

21 Yang, *Luoyang qielan ji jiaojian*, 53. Cf. Wang, *A Record of Buddhist Monasteries in Lo-yang*, 55.

22 I am grateful to Natalie Gummer and Heather Blair for pointing out this resemblance. On pagodas in this period, see Wei-Cheng Lin, "Performing Center in a Vertical Rise: Multilevel Pagodas in China's Medieval Period," *Ars Orientalis* 46 (2016): 100–135.

23 Yang, *Luoyang qielan ji jiaojian*, 53–54; cf. Wang, *A Record of Buddhist Monasteries in Lo-yang*, 55.

24 Gan Bao 干寶, *Sou shen ji* 搜神記, annotated by Wang Shaoying 汪紹楹 (Beijing: Zhonghua Shuju, 1979), 217–18.

25 Berthold Laufer, *Sino-Iranica: Chinese Contributions to the History of Civilization in Ancient Iran, with Special Reference to the History of Cultivated Plants and Products* (Chicago: Field Museum of Natural History, 1919), 221.

26 Robert N. Spengler III, *Fruit from the Sands: The Silk Road Origins of the Foods We Eat* (Berkeley: University of California Press, 2019), 195.

27 Laufer, "Sino-Iranica," 282–84.

28 Jujubes grow to about three centimeters, which is also the size of most grapes purchased in supermarkets today. This suggests that grapes typically encountered in medieval China were much smaller—and serves as a reminder that our own table grapes have been bred for size.

29 Yang, *Luoyang qielan ji jiaojian*, 171–72; cf. Wang, *A Record of Buddhist Monasteries in Lo-yang*, 174–75.

30 Yang, *Luoyang qielan ji jiaojian*, 12; cf. Wang, *A Record of Buddhist Monasteries in Lo-yang*, 17.

31 Yang, *Luoyang qielan ji jiaojian*, 50–51; cf. Wang, *A Record of Buddhist Monasteries in Lo-yang*, 51.

32 D. C. Lau, trans., *Mencius* (Harmondsworth: Penguin Books, 1970), 51.

33 Yang, *Luoyang qielan ji jiaojian*, 143; cf. Wang, *A Record of Buddhist Monasteries in Lo-yang*, 63.

34 Yang, *Luoyang qielan ji jiaojian*, 60. Luoyang would seem to be filled with fruit trees: not too far from this very temple was the "Garden of One Hundred Fruits." It contained the "jujubes of the immortals," which were large and tasty. There were also "peaches of the immortals," whose red color permeated skin and flesh and, like the jujubes, were said to have originated in the sacred Kunlun Mountains 崑崙山. Yang, *Luoyang qielan ji jiaojian*, 63; cf. Wang, *A Record of Buddhist Monasteries in Lo-yang*, 63.

35 Yang, *Luoyang qielan ji jiaojian*, 135; cf. Wang, *A Record of Buddhist Monasteries in Lo-yang*, 138–39.

36 Yang, *Luoyang qielan ji jiaojian*, 12; cf. Wang, *A Record of Buddhist Monasteries in Lo-yang*, 16.

37 Yang, *Luoyang qielan ji jiaojian*, 77. Based on a text identifying the plants and animals in the "Odes of Mao" (a portion of the Classic of Poetry 詩經), Yi-t'ung Wang identifies these as evergreens, hollies, water lilies, and mallows,

but the more descriptive names here are a reminder that plants are hard to identify and nomenclature shifts: what is called a "beef-tendon tree" can now be connected with two different plants. Wang, *A Record of Buddhist Monasteries in Lo-yang*, 48–49.

38 Yang, *Luoyang qielan ji jiaojian*, 124; cf. Wang, *A Record of Buddhist Monasteries in Lo-yang*, 125.

39 Yang, *Luoyang qielan ji jiaojian*, 200; cf. Wang, *A Record of Buddhist Monasteries in Lo-yang*, 204.

40 Yang, *Luoyang qielan ji jiaojian*, 209; cf. Wang, *A Record of Buddhist Monasteries in Lo-yang*, 213.

41 This is true for most bamboo, although there are some species that will lose their leaves during cold weather.

42 For one discussion of the "three friends," see Maggie Bickford, *Ink Plum: The Making of a Chinese Scholar-Painting Genre* (Cambridge: Cambridge University Press, 1996), 52–53.

43 Yang, *Luoyang qielan ji jiaojian*, 60; cf. Wang, *A Record of Buddhist Monasteries in Lo-yang*, 59.

44 Like the scholar-tree above, the Linnaean name of this pollia reflects the fact that Westerners first encountered the plant in Japan.

45 Yang, *Luoyang qielan ji jiaojian*, 88; cf. Wang, *A Record of Buddhist Monasteries in Lo-yang*, 85.

46 Yang, *Luoyang qielan ji jiaojian*, 199; cf. Wang, *A Record of Buddhist Monasteries in Lo-yang*, 203.

47 Yang, *Luoyang qielan ji jiaojian*, 174; cf. Wang, *A Record of Buddhist Monasteries in Lo-yang*, 177.

48 Yang, *Luoyang qielan ji jiaojian*, 179–80; cf. Wang, *A Record of Buddhist Monasteries in Lo-yang*, 196. The Prince of Liang, Liu Wu 劉武 (ca. 184–144 BCE), was the second son of Emperor Wen 文 of the Han. His garden was renowned and said to be the first rock garden in China. See Frances Ya-sing Tsu, *Landscape Design in Chinese Gardens* (New York: McGraw-Hill, 1988), 28.

14

PLANT HUMANITIES JOIN ETHNOBOTANY TO REDUCE CULTURAL PARALLAX AND MORE FULLY PROTECT SACRED PLANTS?

GARY PAUL NABHAN

The fields of plant humanities and ethnobotany often have common interests but divergent methodologies and audiences. Ethnobotany has largely drawn on the methodologies of anthropology and economic botany over the last thirty years, while plant humanities has recently innovated many eclectic methodologies, including but not limited to critical theory, visual and historical content analyses of material objects associated with plant exploration, and greater emphasis on unpacking the influence of class and race on how colonial explorers went about their botanical work. At the same time, the field inquiries of ethnobotanists in communities and their biocultural landscapes have reminded plant humanities scholars of the distortions inherent in the phenomenon of "cultural parallax" that comes from a scholar's cultural distance from the embodied knowledge in the community they are analyzing.[1] Ironically, since my 1994 coining of the term "cultural parallax" in a seminal meeting of humanities scholars and conservation scientists, it has been used three times as often in environmental humanities articles and books than in the scientific literature of ethnobiology and ethnoecology cited in Google Scholar.

Nevertheless, there are fruitful opportunities for blending the methodologies of these two disciplines or at least presenting both to highlight the cognitive dissonance between them. In this case study of the political, cultural, and ethical dimensions of a sacred plant, I wish to illustrate two significant elements of cognitive dissonance between plant humanities and ethnobotany at this moment in time. As an interdisciplinary field shared between the natural and cultural sciences, ethnobotany tends to take a secular, materialistic approach to the cultural dimensions of sacred plants and has often demonstrated an aversion to considering any theological or even ethical interpretations of these relationships. Scholars in the plant humanities have little or no qualms about exploring these dimensions, but humanities scholars are seldom initially conversant with the details of the interactions in question.

The teaming up of ethnobotanists, plant humanities scholars, and Indigenous or other spiritual leaders can play a unifying role in advancing the ethical concerns in Indigenous communities where sacred plants and the traditional ecological and spiritual knowledge about them are being lost at unprecedented rates. It can also give sacred plants the legal standing they deserve.[2] The example of the saguaro cactus as an imperiled sacred plant illustrates that such collaborations can, in fact, lead to legal protection, restorative justice, and ecological restoration of sacred plants (Figure 14.1).[3]

We live by plants, because of plants, and in exchange with plants, some sacred and some profane. But our interactions with plants are by no means static; over the course of history, some sacred plant species have been subject to such demand that they have been overharvested, profanely commodified, or globally endangered, despite the reverence for them. In other cases, war, climate catastrophes, insects, or diseases have diminished access to wild plants, so much so that they have become endangered, protected, or cultivated. With few exceptions, plants considered to be sacred, holy, or of ceremonial and ritual significance are regarded in conservation sciences as entities that deserve protection or recovery simply because of their environmental, genetic, or economic value. Their roles as part of the intangible heritage of cultures and faiths from around the planet are incidental to most biologists. Yet I sense that when sacred and ceremonial plants suffer or become culturally appropriated, something in the human spirit is diminished as well.

But if there is any hope that we can reawaken the divergent cultures and values of science and faith (or the humanities writ large) to rediscover synergies among them, it may be that the field of plant humanities is the ticket. Many scholars have offered their own valuable perspectives on C. P. Snow's *The Two Cultures and the Scientific Revolution*

FIGURE 14.1 Linen postcard, "Sentinels of the Desert," 1931. William L. Bird Postcard Collection, National Museum of American History, Smithsonian Institution, Washington, D.C.

(1959), but it remains true that we need more tangible, innovative collaborations between humanities scholars and scientists to put this false dichotomy to rest.[4] Why not use the *biocultural recovery* of iconic plants of inestimable spiritual and scientific significance to bridge the long-standing divide between ecological science and religion? After all, the etymological roots of both ecology—*oikos* and *logos*, "the logic of the earth household and interactions therein"—and religion—*religio/religare*, "to bind together or reconnect"—have uncanny similarities in their original sense of purpose.

Reconnecting nature with culture is the goal of the Sacred Plant Biocultural Recovery Initiative launched in April 2024 at two events that took place at the Harvard Humanities Center in Cambridge and the Takreem Foundation in Boston during the same week. The initiative involves as advisors several spiritual leaders from more than a half dozen Indigenous Nations, as well as practitioners of the Muslim, Druze, Jewish, Greek Orthodox, Buddhist, Roman Catholic, and Protestant faiths in tribal governments, colleges, and nonprofits around the world. The initiative aspires to model the healing of the science-humanities divide by getting leaders in the trenches together to transplant trees, sow seeds, restore wounded habitats or war-torn borders; and by involving elders and youth in multigenerational collaborations.

As a case in point, at least forty-five species of sacred and ceremonial plants suffered either temporary or permanent declines during the "Trump wall" era from January 2017 to January 2021. During that period, federal and state laws protecting plants, wildlife habitats, sacred sites, and national parks were overridden or waived by a hyped-up federal emergency when President Trump signed Executive Order 13767. The order directed the Army Corps of Engineers to build a wall with the assistance of the Department of Homeland Security and various subcontractors, few of whom had ever worked in the desert environments spanning the boundary of the United States and Mexico. U.S. agencies built 455 miles

FIGURE 14.2
"Saguaro Trees," in Julius Fröbel, *Seven Years' Travel in Central America, Northern Mexico, and the Far West of the United States* (London: R. Bentley, 1859), 508–9. Smithsonian Institution Libraries, Washington, D.C.

(732 km) of new barriers; 49 miles (79 km) had no barrier to seeds, wildlife, or humans prior to 2017.

The repercussions of this construction project exemplify the role that plant humanities can play in documenting, analyzing, and proposing concrete solutions to assist in the biocultural recovery of any sacred plant, in service to its community of faith-based users. Of all the plant populations of a sacred species that suffered during that recent period, perhaps it was the damage, mutilation, and killing of the iconic giant saguaro cacti (*Carnegiea gigantea*) that received the most publicity and triggered the most human grief. This plant is unique to the binational Sonoran Desert, and while not federally listed as threatened or endangered either in the United States or Mexico, it is protected from illegal harvesting and transport without permits in the state of Arizona and has special status in at least five U.S. national parks and monuments (Figure 14.2).

I will present a summary from a humanities perspective—rather than one from a scientific perspective per se—of the difficulties that have occurred in the interactions among saguaros and Indigenous cultures over the last 150 years. I will then focus on how these challenges prompted Indigenous peoples to reassert their right to access this sacred plant for ceremonial purposes and to devise fresh means of safeguarding the biocultural relations with this plant, ecologically, spiritually, and legally. Next, I will use this example as a springboard for plant humanities scholars who wish to be involved with other sacred plants in the service of Indigenous and other faith-based communities. I want to emphasize from the start that the goal of biocultural recovery is not to "conserve" the saguaro genome as an object or "thing" in a seed bank, botanical garden, or nursery. Instead, it must be *relational* work of safeguarding or renewing the cultural interactions between saguaros and their traditional users.

Saguaro Case Study

As early as 1776, Jesuit missionaries remarked upon the special status that the saguaro (called *ha:sañ* in Piman languages) commands in the O'odham communities of the Sonoran Desert, including the Akimel O'odham (River Pima), the Tohono O'odham (the Desert People formerly known as Papago), and the Hia-Ced O'odham (the In-the-Sand People). O'odham scholars themselves—including Brennan, Juan Smith, Francis Vavages, Ofelia Zepeda, Danny Lopez, Camillus Lopez, and Barnaby V. Lewis—have amply documented from oral histories in their own language how saguaros significantly factor into the creation narratives, material culture, seasonal ceremonial cycle, ethics, and spirituality of the O'odham peoples, whose ancestral homelands span an area twice the size of Connecticut. Even though I have periodically worked in O'odham communities since 1975 and was invited to two saguaro ceremonies and countless harvests by O'odham elders, I am not a tribal member, nor do I have a license to divulge or interpret the esoteric oral religious texts of the O'odham that have been translated and transcribed into English over the last 130 years. All I feel able to report from publications distributed generally from the O'odham themselves is the following. In oral histories, two orphaned O'odham boys went up into the mountains, with one brother turning himself into the little-leafed paloverde tree, while the other sank into the ground and reemerged as a saguaro. In one oral history collected by José Luis Brennan, the latter boy declared, "I will turn myself into a saguaro, so that I may last forever, and bear fruit each summer." This positioned saguaros to be regarded as (sacred) sentient beings, and their fruit, when fermented into wine, has become a sacramental element of rain-bringing (*Nawait I'idag*, "sit together and drink saguaro wine") ceremonies at the onset of summer, when O'odham farmers require monsoonal rains and runoff to plant their traditional crops. As such, saguaros have become essential to the O'odham seasonal cycle of ceremonial practice and food production.

The use of saguaro fruit—which requires pollination by nectar-feeding bats, and which is eaten by owls, doves, and many other birds—has spawned a wider range of ceremonial relations. I was personally asked once by a Tohono O'odham elder to drive him out before dawn so that he could leave the first saguaro fruit he collected for a rare ferruginous cactus pygmy owl, to ask permission of it to harvest other fruit to eat and for the rain-bringing ceremonies. "First fruit" rituals are found around

FIGURE 14.3

Alfred Cohn, *Saguaro Fruit Harvest, Sells, Arizona*, ca. 1947. Cabat-Gill Collection, Tucson Historic Preservation Foundation, Tucson.

the world to ask for permission or to offer gratitude to the Creator before other fruit is harvested exclusively for human use.

By 1883, the U.S. Court of Indian Offenses, established by the Secretary of the Interior, attempted to eliminate all Native American spiritual practices to encourage Christianization. Within a decade, this court began to try and prosecute O'odham spiritual leaders presiding over rain-bringing ceremonies. In 1895, 1897, and 1908, law enforcers disrupted ceremonies, then jailed or killed spiritual leaders. However, this harmful intervention against the expression of constitutionally guaranteed religious freedom was overturned during John Collier's tenure as director of the Bureau of Indian Affairs, and a few of the many communities that once practiced *Nawait I'idag* ceremonies resumed them. Since then, no federal, state, or county authority has attempted to extinguish the right of the O'odham to drink the mildly alcoholic saguaro wine. Tribal councils sanctioned the participation of tribal members in these sacred observances.

In 1937, when designating the Organ Pipe Cactus National Monument, President Franklin Delano Roosevelt granted O'odham tribal members the right to harvest cactus fruit there in perpetuity. In internal memos of the U.S. Customs Service (as the Bureau of Customs and Border Protection was then known), the O'odham were granted freedom not only to harvest saguaro fruit in the Roosevelt Reserve—the sixty-foot buffer between Mexico and the United States, along the border fence—but also to freely cross into Mexico to gather fruit. Accordingly, the U.S. government recognized the O'odham right, as binational

FIGURE 14.4

Saguaro Harvest, mural by Michael Chiago. Photograph by Paul Mirocha, 2023, reproduced courtesy of Michael Chiago.

citizens, to harvest fruit even in an area along the border that was legally out of National Park Service control (Figures 14.3 and 14.4).

These historic precedents—largely uncovered through plant humanities research methods and not through the scientific literature of botany—establish theologically and legally that the saguaro cactus fruit and the O'odham people have had a sacred relationship since time immemorial, and that perhaps this relationship should be regarded as *entheogenic* or "God-manifesting" as well.[5] These precedents can then be used to interpret what happened between 2017 and 2021 in Organ Pipe Cactus National Monument that set the stage for the Sacred Plant Biocultural Recovery Initiative.

Soon after the construction of Trump's wall began in southern Arizona, I was contacted by O'odham activist Floyd Flores, who expressed concern about the number of saguaros that were being massacred in O'odham aboriginal homelands; these saguaros were "off-reservation," but still in places on federal lands where the O'odham have rights to practice their spiritual traditions. My wife and I soon invited Indigenous elders with affiliations to five border tribes to join us as a work group that we called Healing the Border Disorder. After a rally against the border wall held at Organ Pipe Cactus National Monument, at which I was asked to speak, we accompanied many O'odham in driving along the border to see firsthand the carnage being done by the Army Corps of Engineers just north of the international border. As I scrambled out of a car at the first stop, where we were guided by National Park Service law enforcers, I saw a middle-aged O'odham woman lying next to a toppled, mangled saguaro. She was weeping, crying out loud: "Don't they know they are people too?" Thousands of saguaros had been destroyed, hauled away to allow for the construction of the border wall. As the *Los Angeles Times* was soon to question in a headline of an article by Molly Hennessey-Fiske on February 26, 2020: "It's illegal to destroy saguaro cactuses. So why are they being removed for Trump's border wall?" (Figure 14.5).

Within a few weeks, I was able to meet with Organ Pipe's superintendent, Scott Stonum, who expressed sympathy for the concerns voiced by the O'odham and the public at large but felt that his hands were largely tied. President Trump had waived fifty-one state and federal laws by his declaration of a "national emergency at the border," and Stonum had convinced the Department of Homeland Security to pay for removing approximately one thousand saguaros intact and transplanting them out of the Roosevelt Reserve into the National Monument, while he used National Park Service funds to transplant at least as many. However, we soon heard reports that Army Corps of Engineers subcontractors working to remove or transplant the saguaros were illegally transporting them out of the monument on state highways in attempts to sell them in Ajo, Arizona, and elsewhere.

The O'odham who witnessed these attempted sales were repulsed and disgusted that anyone would even attempt such an unethical and illegal monetary transaction involving a sacred plant. Our work group filed multiple complaints through the online portals of the Department of Homeland Security, U.S. Customs and Border Protection, Army Corps of Engineers, Mexican Environmental Protection Agency (PROFEPA), and Arizona Department of Agriculture (the latter being responsible for protecting saguaros), stating that federal agencies and their contractors were violating laws regarding saguaros and their "disposal" that had not been and could not be waived. Various members of our work group also wrote op-eds about our concerns for *The Arizona Republic*, *Indian Country Today*, *Arizona Daily Star*, *Franciscan Action*, and *National Catholic Reporter* and were interviewed by at least a dozen other print, radio, and television media. Some of these commentaries underscored the need for government agencies to condone and protect Indigenous rights to sacred plants and places as being equal to—if not much older than—the rights of Western faiths to similar "spiritual resources." There was strong support from interfaith groups and from other Indigenous nations generated from these commentaries.

The next step was for our Healing the Border Disorder work group to assist Tohono O'odham elected officials in drafting a resolution granting "personhood" to saguaros and the legal rights associated with that concept. It also noted that the spiritual use of these desert plants fits the definition of unalienable rights from U.S. Secretary Pompeo's Commission on Unalienable Rights, released in July 2020, in the sense that saguaro fruit is essential to the religious practices of the O'odham engaged in rain-bringing ceremonies and is, therefore, "unalienable cultural property." Resolution No. SXDC 01-21-04 on the recognition and protection of Sacred Hasan (saguaro) was unanimously endorsed by every Tohono O'odham district council and the Nation's legislative council, then signed into law by the Nation's elected officials. It may be only one of three such Indigenous Nation laws in the United States that recommends how to implement such a law in detail, not only on the Tohono O'odham reservation but across the entire aboriginal homelands as well.

While we knew we could not stop the border wall construction itself, we felt that ceremonial activism and prayer at sacred places in its path could raise awareness that President Trump could waive federal laws but not the constitutionally guaranteed right to religious expression, which in our minds included access to sacred plants and spring waters. The day before the border wall was erected at Quitobaquito Springs in Organ Pipe, we organized a prayer vigil and exchange of spring waters carried by youth from desert oases thirty-five miles to the north and thirty-five miles to the south of the border. With both park law enforcement and Border Patrol officers granting permission to Indigenous elders in our group while also remaining present to monitor our activities, more than one hundred of us from seven Indigenous Nations and multiple faiths gathered not in protest but in song and prayer. The event received statewide and national publicity, and the cross-cultural, interfaith exchange had a calming effect on most participants.

As wall construction continued west of the springs over the following weeks, young O'odham activists began to protest road closures and to stage prayer vigils at Quitobaquito without consulting the National Park Service, seeing no need to ask for permission to pray in their own language in a place where their ancestors had done the same over centuries or millennia. On September 6, 2020, Hia-Ced O'odham David and Ortega went out before dawn to Quitobaquito to pray and realized that it had become a major staging site for ongoing construction, replete with bulldozers and other heavy equipment. As the construction crews began to arrive in their vehicles for the day's work, Amber blocked traffic while Nellie Jo sat in the bucket of a front-end loader and refused to leave voluntarily. The women told the construction crew: "It's not okay for you to be here. You do not have permission to be here. This is O'odham land. You do not have permission to take the water; you do not have permission to take our artifacts, our history. You're taking our history; you're trying to erase it."

Both were arrested by National Park Service and Border Patrol enforcers on misdemeanor charges of violating a lawful order of a government officer and violating a closure order. They were chained and held incommunicado in a private prison for twenty-four hours, without the opportunity to call family or lawyers. When their trials approached, David accepted a plea deal for a reduced sentence of a $200 fine and probation, while Ortega insisted that federal law should protect her religious beliefs. In November 2021, Ortega's lawyer argued that she should be protected under the 1993 Religious Freedom Restoration Act, which prohibits federal and state governments from "substantially burdening a person's exercise of religion."

U.S. Magistrate Judge Leslie A. Bowman initially argued that the area's closure did not meet the "substantial burden" requirement because Ortega still had access to Quitobaquito Springs. She also initially declined to hear the testimony of Hia-Ced O'odham elder Lorraine Eiler, who descended from the original inhabitants of Quitobaquito. But in January 2020, Judge Bowman reversed her decision after attorney Amy Knight, Ortega's counsel, argued that Ortega not only had a right to pray but also to protect her ancestral land from destruction. Bowman then allowed Eiler to testify on Ortega's behalf, presenting evidence that O'odham spiritual leaders had

FIGURE 14.5
Dead Saguaro series. The fallen saguaro was photographed every six months until it disappeared into the landscape. Photographs by Paul Mirocha, 2013.

led prayer ceremonies there for decades with the permission of the National Park Service and other agencies, and that the place, as well as its spring water and plant life, was widely regarded as sacred among all O'odham. Judge Bowman immediately dismissed the case after Eiler's testimony, concurring that an undue burden had indeed been placed on those attempting to practice their traditional spirituality in a place recognized as sacred. The settling of this legal case also allowed our Healing the Border Disorder work group to encourage the National Park Service to secure funds to involve O'odham youth—under the direction of tribal elders and monument staff—to begin to restore some of the sacred plants lost during the "Trump wall" era. At the time of this writing, the planning and implementation of this "cultural healing process" continue.

The above discussion begs the question of whether plant humanities and ethnobotany scholars can be in tangible service to Indigenous communities struggling to more fully understand and safeguard sacred and ceremonial plants, as well as their esoteric knowledge about them. The answer for now is a qualified yes. Through the efforts of the Tohono O'odham Nation, assisted by the Healing the Border Project, the Tohono O'odham Legislative Council passed a resolution to halt the removal or damage of sacred saguaro cacti on the aboriginal homelands; to demand restitution payments from perpetrators for tribal professionals who identified, mapped, or restored these plants in biocultural landscapes; and to train contractors and state and federal agency employees through cultural sensitivity workshops on how to do least harm.[6]

In addition, the Department of Defense responded to a formal complaint by the Healing the Border Project by admitting that it had been lax in properly training its own employees and those of its contractors about this issue and agreeing to step up the depth of its cultural sensitivity efforts. Finally, in the fall of 2024, Native American leaders affiliated with seven sovereign Indigenous Nations met with National Park Service administrators and nonprofit organization leaders to devise better strategies on all public lands for advancing stronger protection of sacred and ceremonial plants; assuring access for Indigenous spiritual practitioners and caretakers; and reducing the risks of cultural appropriation of sacred and ceremonial plants growing on federally managed public lands.[7] It remains too early to tell how effective these precedents might be in tangibly safeguarding sacred and ceremonial plants in their biocultural landscapes, but some on-ground restoration efforts are now funded by the National Park Service. These efforts collaborate with tribal elders and youth to restore some of the fourteen species of sacred and ceremonial plants reported in the Quitobaquito Springs biocultural landscape, which is considered a traditional cultural landscape by both the National Park Service and seven affiliated tribes.

From Saguaros to a Broader Suite of Sacred Plants

My preliminary estimate is that there are at least forty-five sacred, ceremonial, and ritual plants significant to Indigenous spirituality and folk Catholic traditions along the entire U.S.–Mexico international boundary area. How many were damaged by wall construction has yet to be fully assessed, but proposals to do so are now being entertained by two federal agencies. In another desert region—the war-torn Middle East—there are more than sixty species of sacred, ceremonial, and ritual plants that undergird the spiritual traditions of Muslims, Druze, Eastern Orthodox, Maronite, Catholic, and Jewish faiths. From Ethiopia to Morocco in North Africa, there are "church forests" and "saint's forests" that have received informal but effective protection from Muslim and Christian faith communities for centuries, if not millennia. The vernacular but highly functioning *hima* reserves for such sacred plants and waters are being renewed and celebrated in almost a dozen countries in the Middle East and North Africa. These community-based reserves stand in sharp contrast to many "top-down" protected areas designated by governments.

How can plant humanities scholars contribute to these efforts? First, by meticulously documenting, visually and archivally, and providing more widespread recognition of the value of sacred plants not just for plant conservation but for the persistence of cultural and spiritual traditions. Second, by comparing and contrasting various community-based strategies for protecting the relationships between sacred plants and people. Third, by comparing and contrasting these faith-based strategies with those implemented in a secular manner by governments, NGOs, and international agencies under the United Nations. Lastly, by contextualizing the ethical and spiritual motivations for this kind of conservation, which are so distinctive when we recognize that most conservation funds go only to secular efforts.

Notes

1 Gary Paul Nabhan, "Cultural Parallax in Viewing North American Habitats," in *Reinventing Nature? Responses to Postmodern Deconstruction*, ed. Michael E. Soulé and Gary Lease (Washington, D.C.: Island Press, 1995), 87–101. See also Christine Woyshner and Jessica B. Schocker, "Cultural Parallax and Content Analysis: Images of Black Women in High School History Textbooks," *Theory and Research in Social Education* 43, no. 4 (2015): 441–68.

2 Christopher D. Stone, *Should Trees Have Standing? Law, Morality, and the Environment*, 3rd ed. (New York: Oxford University Press, 2010).

3 Lorraine Marquez Eiler, Kendall Jose, Verlon Jose, Laura Smith Monti, Austin Nunez, Gary Paul Nabhan, Mona Polacca, and Octaviana V. Trujillo, *Saguaros as Sacred Plants: Healing the Border Project* (Tucson: Alianza Indigena Sin Fronteras, 2021); Gary Paul Nabhan and Octaviana Trujillo, "Sacred and Ceremonial Plants on Public Lands," https://www.garynabhan.com/news/2024/12/sacred-and-ceremonial-plants-on-public-lands/; and Tohono O'odham Resolution SXDC 0121-04.

4 C. P. Snow, *The Two Cultures and the Scientific Revolution* (Cambridge: Cambridge University Press, 1959).

5 For the cultural significance of the saguaro cactus, see William L. Bird, *In the Arms of Saguaros: Iconography of the Giant Cactus* (Tucson: University of Arizona Press, 2023).

6 Eiler et al., *Saguaros as Sacred Plants*.

7 Nabhan and Trujillo, "Sacred and Ceremonial Plants on Public Lands."

INDEX

Figures and tables are indicated by "f" and "t" following page numbers.

C

D

E

F

M

Q

R

S

T